D0186745

Environment
Science
Materials
Technology

A

9

000

Other titles of related interest:

Building Quantities Explained, fourth edition
Ivor H. Seeley

Building Technology, fifth edition
Ivor H. Seeley

Construction 1
Management – Finance – Measurement
A.V. Hore, J.G. Kehoe, R. McMullan and M.R. Penton

Estimating, Tendering and Bidding for Construction
Adrian J. Smith

Introduction to Building Services, second edition
E.F. Curd and C.A. Howard

Understanding Hydraulics
Les Hamill

Understanding Structures
Derek Seward

Construction 2

Environment
Science
Materials
Technology

A.V. Hore, J.G. Kehoe,
R. McMullan and M.R. Penton

MACMILLAN

First published 1997 by
MACMILLAN PRESS LTD
Houndmills, Basingstoke, Hampshire RG21 6XS
and London
Companies and representatives
throughout the world

ISBN 0–333–64949–4

A catalogue record for this book is available
from the British Library.

Typeset in Great Britain by
Aarontype Limited
Easton, Bristol

Printed in Hong Kong

Contents

Preface

This book studies the techniques of construction technology and services, the principles of environmental and materials science, and their applications. It also studies the nature and the historical development of the built environment together with the roles of people working in the construction industry. This wide range of topics is of practical use to students and practitioners studying and working in building construction, civil engineering, surveying, planning and development.

The text is intended to help a wide range of students studying construction and built environment topics. The contents will satisfy the principal requirements of courses and self study for GNVQ/A levels, Edexcel/BTEC awards, HND/HNCs, degrees and professional qualifications. The style of writing is kept simple and supported by clear explanations, a structured layout, practical examples and diagrams. The highlighted definitions, checklists and keyword summaries will also help students preparing for tests, examinations and assignments. The text assumes a minimum of prior knowledge and uses underlying principles to develop an understanding of topics used by professional practitioners in the construction industry.

Some subjects have been distilled into complete treatments which have the advantage of being the first of their type and which enable everyone to gain a good overview of the construction landscape. However, all the subjects deserve further investigation and we hope that this book will be a starting point for many further studies and careers.

Alan Hore, Joseph Kehoe, Randall McMullan, Michael Penton

1 Nature of the Environment

The word *environment* refers to general surroundings, and often means the conditions that surround us as people. This can be a very large topic which might start with the environment of a star, such as the Sun, and close in on the air around just one person.

This chapter considers the effects that humans and their buildings have on the environment. Many of these effects, such as cities spreading into the countryside, are now considered undesirable and are often in the news. Extracting large quantities of materials, such as stone and timber, can leave damaging effects on the countryside unless measures are also taken to sustain and restore the environment.

Natural environment

The natural environment consists of those surroundings which exist without interference from human beings. Notable features of the natural environment include sunshine and rain, mountains and hills, rivers and lakes, rocks and soil, and trees and plants.

Remember that the natural environment can undergo big changes without the presence of human beings. The climates of Britain, for example, have included both tropical periods and ice ages. Mountains have crumbled and been washed to the sea; and you could walk on dry land where the Channel Tunnel now crosses beneath a sea.

Climate

The atmosphere around us produces varying effects of temperature, precipitation (rain, snow and hail), humidity,

1

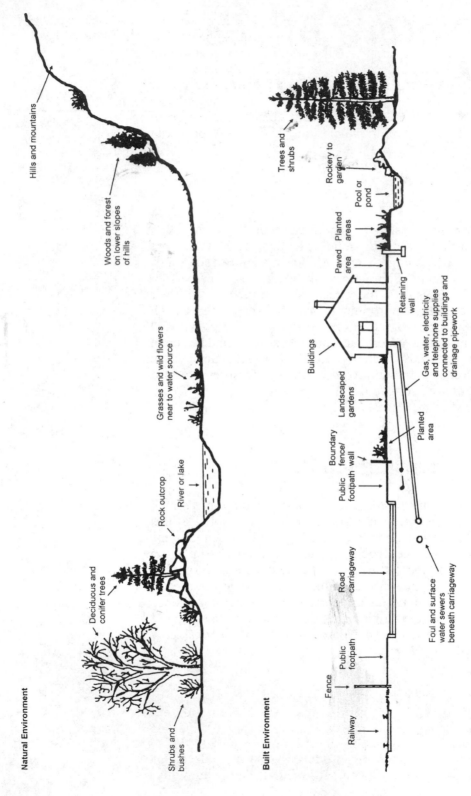

Figure 1.1 Features of the environment.

pressure and winds, cloudiness and visibility. The term *weather* refers to the short-term variations in the atmosphere. The climate of an area is the set of weather conditions typical of that area, and the study of these conditions is known as *climatology*.

Climate affects all human activities, directly or indirectly, as it influences the soils, vegetation and water resources that we depend on for life. The climate of an area determines its type of settlement, shelters, agriculture, manufacturing, transport and other activities in that area.

The climate experienced at a particular place depends largely on the factors described below.

Latitude and season of year
The solar radiation received by an area greatly depends on the angle of the Sun's rays and the length of the day.

Land and sea
Water is slower to heat than land but water stores more heat and is therefore slower to cool. This stored heat helps keep the coastal countryside warmer than inland areas, although the seaboard land may have more rain.

Altitude and topography
The temperature of air decreases as the height above sea level increases. The presence of hills and mountains also produces clouds and rain.

The climate of the British Isles, for example, can be classed as a variable but temperate *maritime* type of climate. There are no great extremes of weather such as the cold winters and hot summers found in the middle of a continent. The main weather systems arrive from the Atlantic Ocean and meet the mountains on the west coasts of Ireland and Britain.

Greenhouse effect

The 'greenhouse' effect is a popular term for the role that the atmosphere plays in keeping the Earth's surface warmer than it might otherwise be. This is a similar mechanism to a garden greenhouse where the glass allows radiation from the Sun to enter but does not allow the heat radiated by the plants to escape. In a similar manner, the inwards radiation from the

Climate
Temperat
Sunshine
Wind
Rain
Snow

Topography features
Mountains
Hills
Valleys
Cliffs
Plains
Streams
Rivers
Lakes
Seas

Ground cover features
Trees
Bushes
Grasses
Crops
Flowers

Sun easily penetrates the atmosphere but is re-radiated from Earth as radiation of a different wavelength, much of which is absorbed by the atmosphere and passed back to the surface.

If there weren't any greenhouse effect then the average ground temperature of the Earth would fall from 15°C to −30°C, as happens on other planets. Obviously we need the greenhouse effect but we don't want to raise the temperature of the Earth. We run the risk of creating an excessive greenhouse effect by increasing the quantity of carbon monoxide, carbon dioxide and other gases in the atmosphere. These gases are given off when we burn fuels such as oil or coal in power stations, or petrol in cars.

Topography

Topography refers to the shape of the land and the features on the land. The positions of hills, plains and rivers affect where we place buildings and how we build them. Topographical maps show this information by shading and by *contour lines* of equal height.

For example, the use of a river for water and for easy transport has been the reason for the location of many early towns, such as London. Land by a river, however, may also have the risk of flooding and provide difficult ground conditions for foundations.

The top of a hill has been the starting point for other towns which needed to be defended. But as we go higher into the hills or mountains it is harder to grow crops and the transport links, such as roads, become more difficult. The different climates between low and high places also affect the type of buildings needed.

Geology

Geology deals with the study of the planet Earth, especially the materials of which it is made and the processes that affect these materials. The term *mineral* can be applied to most of the natural materials, such as rocks, which form the solid Earth. The chemical definitions can vary but minerals have a crystalline structure and are inorganic, which means no living materials are included.

- **Soil** is the natural collection of particles that can be separated into individual particles by relatively gently actions, such as movement in water
- **Rocks** are the solid substances that form the Earth's crust.

A knowledge of soil and rock conditions is needed for the design of buildings and large structures. A *geological map* of an area shows the underlying layers of rocks, although local variations need to be determined by an investigation on site.

Classes of rock

The rocks of the Earth are grouped according to the way that they have been formed:

- **Igneous rocks** solidified from molten material such as magma and lava
- **Sedimentary rocks** formed by the build-up of sediments and particles out of the water or the air
- **Metamorphic rocks** have been transformed by heat, pressure or chemical change.

Geotechnics is a branch of engineering which studies soils so as to design and construct buildings, roads, dams and other structures. Important properties of soil mechanics include the strength, compressibility and permeability (water penetration) of a soil.

Igneous rocks
Basalt
Granite

Sedimentary rocks
Limestone
Dolomite
Shale
Sandstone

Metamorphic rocks
Marble
Schist
Slate
Quartzite

Vegetation

Most of the Earth's surface is naturally covered by trees, plants or grasses which interact with other living organisms and form part of the natural *ecology*.

- **Forests** are areas with numerous trees, plants and animals which depend upon one another for survival
- **Pastures** are open areas with grasses and other vegetation which may be grazed on by wild or farmed animals.

Forests play an important part in natural processes such as absorbing carbon dioxide from the atmosphere, recycling nutrients such as nitrogen, and purifying water. Forests and pastures occur naturally but can also be created, maintained, or destroyed by our activities of forestry and farming.

For thousands of years the landscapes of the British Isles have been altered by human activities. However, where there is a satisfactory balance of forests, fields and other features, this 'countryside' is generally regarded as part of the natural environment rather than the built environment.

Hardwood trees
Ash
Beech
Birch
Cedar
Elm
Mahogany
Meranti
Oak
Walnut

Softwood trees
Douglas Fir
Larch
Parana Pine
Pine
Redwood
Spruce

Built environment

The built environment is formed by the buildings and other objects that humans construct in the natural environment. In addition to the buildings in which we live and work, the built environment includes the *infrastructure* which services and connects these buildings.

Buildings and towns are connected by roads, railways and communication links which in turn need bridges, tunnels, towers and cables. The built environment also includes the various features of the water supplies and electricity supplies needed by our buildings.

Features of the built environment are summarised below, and the items listed form the focus of study for many chapters in this book.

Features of the built environment

Buildings:
    dwellings
    shops and offices
    factories
    public buildings
	Walls
	Fences
	Paths
	Driveways
	Gardens

Power stations Transmission lines Cables Radio/TV masts Telecommunications Pipelines Sewers Sewage works	Roads Railway Seaports Airports Canals
Mines Slag heaps Gravel pits Clay pits Quarries Refuse sites Landfill	Bridges Tunnels Dams Weirs Aqueducts

Preserving the environment

We need to manage our use of the environment so that it gives benefit to current generations of people but also maintains the potential of the environment for future generations. Forests, soils, supplies of fuel and water are examples of resources which need to be managed.

Also important are the interrelationships between the environment, use of resources and living things, which are studied under the title of *ecology*. For example animals, including humans, use oxygen to breath and the amount of oxygen in the atmosphere is influenced by forests and jungles which replace this oxygen.

The use of one resource, such as removing trees from a forest, has effects on other resources such as soil erosion, loss of water supplies and loss of wildlife.

Conservation

Conservation is concerned with managing the environment to make sure there are adequate supplies of natural resources for everyone, now and in the future.

- **Renewable resources** are resources such as forests which, if managed correctly, can regenerate or even improve their resource value.

- **Non-renewable resources** are resources such as oil which exist in fixed amounts on the Earth and, when used, do not regenerate

Energy conservation

Buildings and their services typically use 40 to 50 per cent of all the energy used in a modern country and about half of this building energy is used in domestic buildings such as houses. Therefore, changes in the design and use of buildings can have a large effect on the consumption of energy in a country and in the world.

Most of the energy used to heat buildings, including electrical energy, comes from non-renewable fossil fuels such as oil and coal. This energy originally came from the Sun and was used in the growth of plants such as trees. Then, because of changes in the Earth's geology, those ancient forests eventually became a coal seam, an oil field or a natural gas field. The existing stocks of fossil fuels on Earth cannot be replaced and, unless conserved, they will eventually run out

Methods of conserving energy in buildings are influenced by the costs involved and, in turn, these costs vary with the types of buildings and the current economic conditions. Some of the important options for energy conservation in buildings are described overleaf.

Renewable resources
Soils
Inland waters
Plants
Trees

Non-renewable resources
Oil
Natural gas
Coal
Metal ores
Rare stone

Energy is the capacity of a system to do work, such as moving against a force.

Types of energy
Mechanical energy:
    potential
    kinetic
Heat energy
Electrical energy
Chemical energy
Nuclear energy

Fuels are sources of energy.

Alternative energy techniques

Solar energy

All buildings gain some casual heat from the Sun during winter but more use can be made of solar energy by the design of the building and its services. Despite the high latitude and variable weather of countries in North Western Europe, such as the United Kingdom, there is considerable scope for using solar energy to reduce the energy demands of buildings.

The utilisation of solar energy needn't depend on the use of special 'active' equipment such as heat pumps. Passive solar design is a general technique which makes use of the conventional elements of a building to perform the collection, storage and distribution of solar energy. For example, the afternoon heat in a glass conservatory attached to a house can be stored by the thermal capacity of concrete or brick walls and floors. When this heat is given off in the cool of the evening it can be circulated into the house by natural convection of the air.

Natural sources

Large amounts of energy are contained in the Earth's weather system, which is driven by the Sun, in the oceans, and in heat from the Earth's interior which is caused by radioactivity in rocks. This energy is widely available at no cost except for the installation and running of conversion equipment. Devices in use include electricity generators driven by windmills, wave motion and geothermal steam.

Energy efficiency

The total energy of the Universe always remains constant but when we convert energy from one form to another some of the energy is effectively lost to use by the conversion process. For example, hot gases must be allowed to go up the chimney flue when a boiler converts the chemical energy stored in a fuel into heat energy. Around 90 per cent of the electrical energy used by a traditional light bulb is wasted as heat rather than converted to light.

Efficient equipment

New techniques are being used to improve the conversion efficiency of devices used for services within buildings. Condensing boilers, for example, recover much of the latent heat from flue gases before they are released. More efficient

(*continued*)

Alternative energy techniques (*continued*)

forms of electric lamp, such as fluorescent, are available. Heat pumps can make use of low-temperature heat sources, such as waste air, which have been ignored in the past.

Electricity use

Although electrical appliances have a high energy efficiency at the point of use, the overall efficiency of the electrical system is greatly reduced by the energy inefficiency of large power stations built at remote locations. It is sometimes possible to make use of this waste heat from power stations for various uses in industry and for the heating of buildings using techniques of CHP (Combined Heat and Power).

In the absence of CHP, savings in national energy resources are made if space heating is supplied by burning fossil fuels, such as coal or gas, inside a building that is to be heated rather than by burning those fossil fuels at a power station. Electrical energy will still be required for devices such as lights, motors and electronics but need not be used for heating.

Thermal insulation

External walls, windows, roof and floors are the largest areas of heat loss from a building and standards of thermal insulation in the United Kingdom, as defined by Building Regulations, have scope for continued improvement.

The upgrading of insulation in existing buildings can be achieved by techniques of roof insulation, cavity fill, double-glazing, internal wall-lining and exterior wall-cladding. Increasing the insulation without disrupting the look of a building is a particular challenge for countries with stocks of traditional buildings.

Ventilation

The warm air released from a building contains valuable heat energy, even if the air is considered 'stale' for ventilation purposes. The heat lost during the opening of doors or windows becomes a significant area of energy conservation, especially when the cladding of buildings is insulated to high standards.

These ventilation losses are reduced by better seals in the construction of the buildings, by air-sealed door lobbies, and the use of controlled ventilation. Some of the heat contained in exhausted air can be recovered by heat exchange techniques such as heat pumps.

Threats to the environment

The following sections outline some effects which human activities have on the state of the natural environment. Most of these activities, such as use of energy for buildings and for transport, are linked with the built environment.

Global warming

Fossil fuels
Coal
Oil
Petrol
Diesel
Gas

During the long history of the Earth there have been some large changes in climate, probably caused by variations in solar radiation, volcanic eruptions, impacts of large meteors and other events. Many of these causes are beyond human control but there is evidence that human activities can now cause undesirable changes in the climate.

The atmosphere of the Earth provides a *greenhouse effect* which allows radiant heat into the Earth but prevents all heat from escaping. This warming effect is increased by the burning of fossil fuels, such as in power stations and in motor vehicles. The reduction of the *ozone* layer in the atmosphere by the escape of fluorocarbons can also contribute to the warming of the atmosphere.

An increase in average world weather temperatures causes concern because of resulting rises in sea levels and flooding of lowland, increases in desert area, and difficulties with food crops.

Loss of resources

Loss of resources
Forests
Agricultural land
Variety of plant types
Unpolluted water
Energy resources

Some of the world's resources are non-renewable. For example, coal and oil were formed over millions of years and, once used, cannot be replaced; except after a very long wait!

Pollution

Pollution
Water pollution
Soil erosion
Solid waste
Pesticides
Noise pollution
Air pollution

Discharge of materials or energy into land, water or the atmosphere is detrimental to the ecological balance of the Earth, or can lower the quality of life – a process known as ***pollution***.

Problem areas

Table 1.1 summarises some current activities and consequences which can cause environmental problems. Many issues, such as the use of transport, raise matters of national policy which affect us all.

Table 1.1 Table of environmental problem issues

Areas of activity	Some associated problems
Use of buildings	Loss of natural countryside for building sites. Consumption of energy for heat, washing, cooking, lighting. Use of non-renewable energy such as oil, natural gas.
Extraction of materials	Loss of natural countryside. Annoyance, noise and pollution caused by use of machinery and heavy transport. Unsightly quarries, waste heaps. Instability of ground after mining.
Construction operations	Processing of materials can lead to ground and air pollution.
Transport industries	Loss of natural countryside for road, rail, airports, and associated land. Use of non-renewable energy such as oil. Exhaust gases contribute to poor quality air and global warming.
Agriculture	Loss of natural countryside for industrial farming methods. Use of chemicals which enter the food chain and the water supply. Use of heavy machinery and heavy transport.
Manufacturing	Loss of natural countryside for greenfield sites. Large-scale use of fuels contribute to global warming and to poor quality air. Use of heavy road transport for movement of materials.
Retail activities	Loss of natural countryside to make large out-of-town shopping centres. Transport needed to reach centre sites uses non-renewable fuels; contributes to poor quality air; contributes to global warming.

Preservation of the environment

Table 1.2 summarises some methods and techniques which are environmentally friendly.

Sustainable development is a technique where a resource is managed so that it is replaced at the same rate as it is used. For example, if the trees in a particular forest take 20 years to grow then one-twentieth of the forest could be used each year, provided that replacement trees are planted.

Table 1.2   Good environmental techniques

Technique	Examples
Reuse of materials	Use of stone, brick, timber, fireplaces from earlier buildings.
Recycling of materials via a process	Processing of old paper, metals, concrete, plastic to make new batches.
Replacement of resources after use	Replanting forests.
Preservation of habitats	Keeping old trees on new building sites. Keeping meadows and wetlands for their wildlife.
Reinstatement of land after use	Filling quarries, replanting vegetation.
Recycling of land	Reusing the land of old factories.
Low energy devices	Solar energy. Efficient boilers. Efficient lighting.
Low pollution techniques	Filtered chimneys. Containment of fine particles and liquids. Quietened machines.

Key words for Chapter 1, Nature of the Environment

The following is a list of some keywords used in this chapter. Use the list to test your knowledge and, if necessary, consult the text to learn about the terms.

Climatology	Greenhouse effect	Minerals
Conservation	Hardwood trees	Non-renewable
Fossil fuels	Infrastructure	resources
Geotechnics	Latitude	Sustainable
Global warming	Metamorphic rocks	development
		Topography

2 Development of the Built Environment

The previous chapter described the features of the natural environment, the built environment, and the challenge of creating a built environment without destroying the natural environment.

This chapter looks at the built environment and the factors that help shape it. Features of buildings from the past are described and the factors that affect the design and development of modern buildings are explained. The examples for this chapter tend to be from the British Isles but the principles are universal.

Early built environment

Some early architectures
Egyptian
Mesopotamian
Greek
Roman

The early human occupation of Britain produced some effects which can still be seen. Examples include ditches, hill forts, stone houses, field outlines, tracks and roads.

Those early buildings which survive, such as pyramids, stone circles and cathedrals, are often official or religious buildings rather than houses. Parts of old transport systems such as wharves, canals and roads also survive for many years. We can still see old Roman roads in Britain and the shapes we make across the countryside for railways and motorways will remain long after us.

Materials and techniques

The early houses of a town reflect the materials that were available in that district. For example, wooden weather boards were used for houses near the forests of Kent, Cotswold stone in the West of England, and bricks in areas of clay.

Stone

A simple use of stone is to arrange slabs of stone. Small stones or rubble can be cemented together to make walls by using mortar made from crushed limestone or chalk.

Blocks of stone, smoothly cut or 'dressed', can be arranged to produce more complex structures such as the arch or the dome.

Timber

Two lengths of timber leaning together make a simple frame for the end of a tent. A development of this structure is the *cruck frame* which can be found in medieval buildings such as barns and houses, along with other timbering.

Timber has been used to make both the frames of buildings and the cladding attached to the frames, such as weather boarding.

Clay

Cob is a simple building material produced from wet clay, often with straw added. The wet mixture was heaped and pressed into shape to make walls and then allowed to dry naturally. The cob was usually protected by a coat of plaster.

The common clay bricks and tiles that we still use are made from clay which is 'fired' in a hot kiln. The techniques for stone structures, such as arches and domes, can also be used in brick.

Plaster

Plaster is a material that is coated wet on to walls, ceilings, or into cracks. Traditional plasters are made from the relatively common rocks of limestone/chalk or gypsum.

Styles of building

The manner in which materials are arranged in the buildings of an area produces visible differences or architectural 'styles'. The architecture of earlier buildings deserves a book to itself but notable features of important styles in Britain are summarised below. Most people in Britain live in or near to towns in which there are some examples of most architectural styles.

Figure 2.1   Timber cruck frame house.

Local building materials
Rubble stone
Cut stone
Lime mortar
Green timber
Seasoned timber
Clay cob
Clay brick
Clay tiles
Straw thatch
Lime plaster
Gypsum plaster

Some architectural styles
Saxon
Norman
Gothic
Tudor
Jacobean
Georgian
Victorian
Modern
Contemporary

Medieval

Medieval is a general term for houses and farm buildings constructed during the period when larger buildings might be 'Norman' or 'Gothic' styles, as described below:

- Dates: 1000 to 1500 approximately
- Examples: old barns, farmhouses, manor houses, towns such as York, Chester
- Typical features: timber frames, frame spaces filled with other materials.

Norman

- Dates: 1050 to 1200 approximately
- Examples: Tower of London, early castles, early churches, Durham cathedral
- Typical features: semi-circular arches on large round columns, massive stonework, infilled with rubble.

The term *Norman* is taken from the kings who came to England in 1066 from Normandy in France. The style is also known as *Romanesque*.

Gothic

- Dates: 1150 to 1500 approximately
- Examples: churches, cathedrals
- Typical features: pointed arches, decorated stonework, increasing use of stone tracery and windows.

This style was also copied as 'mock Gothic' during the Victorian period.

Tudor, Elizabethan, Jacobean

- Dates: 1500 to 1620 approximately
- Examples: Hampton Court Palace, towns such as Stratford-upon-Avon
- Typical features: leaded windows, wood-panelled rooms, elaborate brick chimneys, overhanging floors called jetties.

Norman

Gothic

Tudor

Georgian

Victorian

20th Century

Figure 2.2 Architectural styles.

Stuart, Classical

- Dates: 1620 to 1720 approximately
- Examples: Whitehall Banqueting Hall by Inigo Jones, Saint Paul's Cathedral by Christopher Wren
- Typical features: designs taken from Greek and Italian forms, regular patterns, windows, tile hanging on walls, steep roofs with dormer windows. The Great Fire of London (1666) was an occasion for rebuilding London in new styles.

Georgian and Regency

- Dates: 1720 to 1840 approximately
- Examples: London churches by Nicholas Hawksmoor; Houses by Robert Adams; buildings and houses in Bath and Edinburgh; London terraces by John Nash; canals and locks; bridges by Thomas Telford; landscape gardens by Capability Brown
- Typical features: stucco and plaster covering brickwork, cast-iron balconies and railings, symmetrical design, windows with six or eight panes, bow windows, slate roofs, first industrial buildings, first iron structures, first modern cement.

Victorian

- Dates: 1840 to 1900 approximately
- Examples: terraces of houses in cities; railway stations such as St Pancras; Crystal Palace; Houses of Parliament; Royal Albert Hall; town halls; railway tunnels and bridges; shipping docks; water and sewerage systems
- Typical features: variety of styles in stone, red brick, terra cotta; revival of pointed Gothic style; use of cast iron.

Twentieth century

- Dates: 1900 to present
- Examples: garden cities, tower blocks of flats; modern housing estates; office towers; commercial and retail estates; electricity grid; masts for radio, TV and telecommunications; motorway systems; airports, suspension bridges
- Typical features: reinforced concrete, glass, steel cables, modern plastic materials, composite materials, tall buildings, *Art Deco* (1930s) style, modern (undecorated) style.

Some Engineers
Thomas Telford (1757–1834)
John McAdam (1756–1836)
Henry Darcy (1803–1858)
Isambard Brunel (1806–1859)
Ferdinand de Lesseps
    (1805–1894)
Benjamin Baker (1840–1907)
Alexandre Eiffel (1832–1923)
George Geothals (1858–1928)
Othmar Ammann
    (1879–1965)

**Some planners/
    architects/builders**
Inigo Jones (1573–1652)
Christopher Wren
    (1632–1723)
John Vanbrugh (1664–1726)
Robert Adams (1728–1792)
John Nash (1752–1835)
George Gilbert Scott
    (1811–1878)
Richard Norman Shaw
    (1831–1912)
Edwin Lutyens (1869–1944)
Frank Lloyd Wright
    (1869–1959)
Le Corbusier (1887–1965)

Garden cities
Hampstead Garden Suburb
Welwyn Garden City
Port Sunlight

Suspension bridges
Humber (Hull)
Verrazano-Narrows
    (New York)
Golden Gate (San Francisco)
Bosphorus (Istanbul)

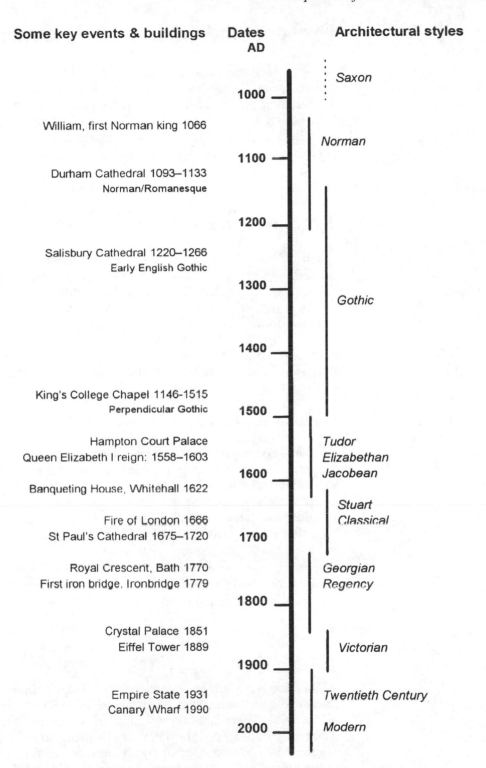

Figure 2.3 Timelines for building periods.

Present and future environment

We wish to preserve good features from our past environment and to promote good features in the new environment that we are building. There needs to be a balance between the various factors and interests which affect the environment, so various guidelines and laws help us to maintain a safe and satisfying environment.

Types of buildings

The buildings of the modern built environment are designed and built for different purposes and these purposes, in turn, affect the features of each building and the facilities that it provides.

The many different buildings in a town or city can be grouped into the following broad classes based on what activities take place in the building. Such *zones* of use are part of town planning principles which help to ensure that houses and factories, for example, are not mixed together in the same street or area.

- **Residential buildings** such as houses, flats, hotels, hostels, hospitals, nursing homes
- **Non-residential and leisure buildings** such as schools, libraries, churches, sports centres, museums, cinemas
- **Commercial buildings** such as shops, banks, restaurants and other places open to the public
- **Business and industrial buildings** such as offices and laboratories, warehouses and factories.

The official grouping of a building also affects regulations, such as those for building and fire, the taxes to be paid, and other laws linked to that group of buildings.

Dwellings

A *dwelling* is a place where people live. Your dwelling or residence is the place where you usually sleep and keep all your belongings. There are many different styles of dwellings for different groups of people. They can be grouped by the way they are joined to one another or by the number of floors or storeys in the building. Common terms used in the United Kingdom are given on page 23.

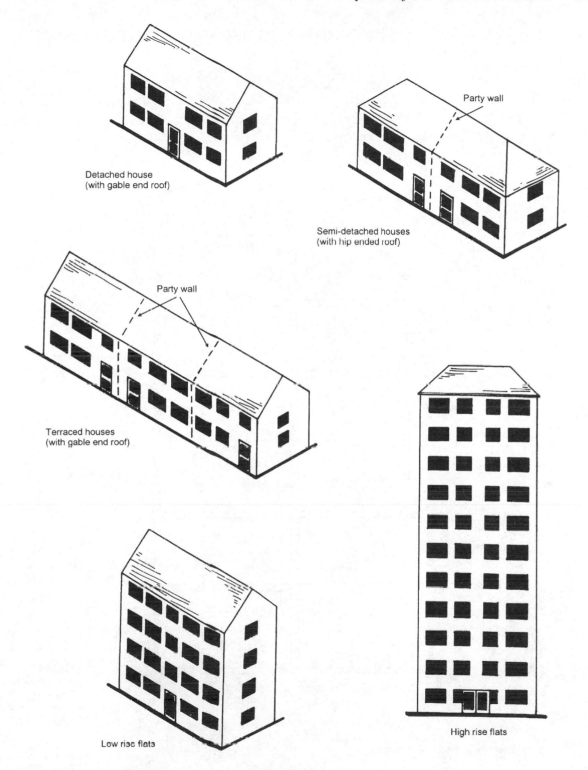

Figure 2.4 Grouping of buildings.

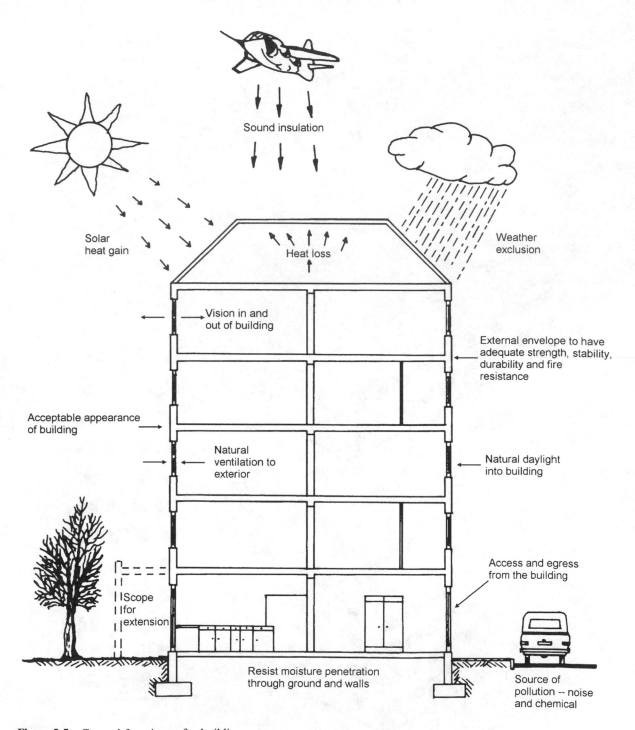

Figure 2.5   General functions of a building.

Types of dwelling

- **Bungalow:** a single-storey dwelling
- **Chalet:** a bungalow with rooms in the pitched roof
- **House:** a dwelling with more than one storey
- **Flat:** a dwelling on one floor in a multi-storey block of similar flats
- **Maisonette:** a dwelling on more than one floor in a multi-storey block of similar dwellings.

Groupings of buildings
Detached
Semi-detached
Terraced
Low rise flats
High rise flats

Development of buildings

Early buildings tended to grow and evolve to suit the needs of those who built them. Materials from nearby were used in a way that was simple for the local builders. These two factors still dominate the resources needed to produce a building:

- availablity of materials
- technical ability of the builders.

Table 2.1   Development aims and constraints

Development aims	*Development constraints*
Suitable appearance Capital costs and running costs Technical performance when finished	**Financial factors** Funds available Interest rates Profit margins National economy
	Legal factors Ownership of land Planning controls Building controls Community laws Contracts between parties
	Social factors Preservation of historical monuments Preservation of trees Conservation of countryside Transport policies Noise pollution Appearance

General constraints
Time allowed
Quality required
Funds available

The development and design of a modern building also involves balancing the various needs, abilities and limitations of: the client who will use the building; the building team who will make it; and the community surrounding the building. The development of a typical building involves balancing the performance targets of the building against any restrictions or constraints.

Performance is the ability of a building and its components to carry out their planned purpose.

*A **constraint** is something which restricts the design or construction of a building.*

Typical factors which affect the aims and performance of buildings or impose contraints are summarised in Tables 2.1 and 2.2

Table 2.2   Performance aims and constraints

Performance aims	*Technical constraints*
Stability	Construction techniques
Strength	Ability of team
Appearance	Season of year
Durability	Location of site
Weather exclusion	Surrounding buildings
Thermal comfort	Ground conditions
Lighting	Access to services
Ventilation	
Noise control	
Sanitation	
Security	

Development of the built environment

The following general conditions can create the desire or the need to develop the built environment:

- **Population movements:** people need houses and other facilities if they move around the country, such as to a new town with new jobs
- **New communication links:** new or improved roads, railways and airports attract people to work and to live.

Table 2.3 Development issues

Some issues	*Typical regulations and constraints*
Population movements	Zoning
Private property rights	Protected buildings
Traffic flows	legislation
Air quality	Protection of trees legislation
Noise pollution	Town and Country Planning Act
Visual pollution	Control of Pollution Act
Refuse handling	Building Regulations
Sewage provision	Health and Safety at Work Act
New buildings: type, size, style	(HASWA)
Alternative energy sources	Noise at Work Regulations
Energy conservation	(NAWR)
Chemical pollution	Factories Acts
Rainforest depletion	CDM (Construction Design and
Global warming	Management) Regulations
Ozone depletion	
Recycling	

Some of the issues associated with development of the built environment are summarised in Table 2.3. Many of these areas have links with the environmental issues described in Chapter 1.

Key words for Chapter 2, Development of the Built Environment

The following is a list of some keywords used in this chapter. Use the list to test your knowledge and, if necessary, consult the text to learn about the terms.

Clay cob	Georgian style	Christopher Wren
Constraint	Lime plaster	Thomas Telford
Cruck frame	Norman style	
Dwelling	Tudor style	

3 *Organisations and Careers*

Environmental challenges
Alternative energy
Alternative materials
Alternative techniques
Sustainable use
Minimal pollution
Minimal noise
Protection of habitat

Construction and other activities in the built environment are a major industry and employer in any country. In the United Kingdom, for example, the yearly value of construction works is approximately 10 per cent of the gross national product and over one million people are employed in the industry.

All around us in the built environment you can see a great variety of construction projects. They range from house repairs and extensions to large tunnels and bridges. In an established country like Britain, the renovation of existing sites and the restoration of existing buildings are an additional challenge.

The design, production and care of the built environment will always offer a wide choice of jobs. The technologies, such as energy conservation, used in modern construction are increasingly complex and need qualified people.

Health and safety matters are always of concern for people working in the built environment. The construction of buildings and structures involves some operations with potential hazards unless they are carried out correctly. There is also a wider awareness of 'green' issues associated with the use of our built environment, and these have brought new challenges and solutions. Chapter 1 of this book describes the issues and challenges of preserving the environment.

This chapter looks at the various occupations available in the built environment sector (see Table 3.1). In addition to jobs, the chapter also describes the areas, organisations and qualifications associated with working in the built environment.

Occupations in construction and the built environment

Built environment sectors

The activities of constructing and using the built environment can be divided into the following main areas of work.

Table 3.1   Occupations in construction and the built environment

Architecture, Building, Surveying
architect
architectural technologist
bricklayer
builder
building control officer
building maintenance manager
building surveyor
buyer
carpenter
clerk of works/inspector
contracts manager
draughtsperson
estimator
floor covering installer
glazier
heating and ventilating engineer
joiner
land surveyor
owner/manager
painter and decorator
planner
plant operator
plasterer
project manager
quantity surveyor
roofer
shopfitter
site engineer
site manager
site supervisor
stone mason
tiler
wood machinist

**Town Planning, Housing,
   Property Management**
conservator
estate manager
facilities manager
housing manager
land planner
planner/consultant

planning technician
property surveyor
valuer

Civil and Structural Engineering
buyer
civil engineering consultant
civil engineering technician
estimator
ganger
general foreman
highways maintenance
land surveyor
material technologist
municipal engineer
planner
project manager
quantity surveyor
site engineer
site manager/engineer
site supervisor
steel erector
stone mason
structural design consultant
water engineer

Building Services Engineering
building services designer/
   consultant
building services surveyor
buyer
clerk of work/inspector
electrician
estimator
facilities manager
maintenance manager
owner/manager
planner
plumber
project manager
quantity surveyor
site manager/engineer
site supervisor

Building construction

Building is concerned with the design, construction, maintenance and conversions of buildings.

Built environment teams
Client
Architect
Quantity surveyor
Civil engineer
Structural engineer
Contracts manager

Site manager
Site engineer
Building services engineer
Clerk of works
Building control officer
Building surveyor
Technicians

General foreman/forewoman
Supervisor
Skilled craftsman/
  craftswoman
Semi-skilled operative
Chargehand
Ganger
Operative or labourer
Subcontractors

Local authority technical
  officers
Building control officer
Health and safety inspector

Suppliers
Delivery drivers
Store staff
Personnel staff
Security staff

- *Examples of buildings*: Homes, shops, schools, recreation centres, hospitals, places of worship and assembly, offices, warehouses, factories.

Civil engineering construction

Civil engineering is concerned with the construction and maintenance of large infrastructure items which are usually for public use.

- *Examples of civil engineering structures:* roads, railways, airports, seaports, bridges, tunnels, dams, retaining walls, sea defences, dams, water towers.

Building services engineering

Building services engineering is concerned with the installation, commissioning and maintenance of the services which operate in a building.

- *Examples of services:* plumbing installations, electrical installations, heating and ventilating, refrigeration, lifts, escalators, fire protection, security systems.

Planning, regulation, property development, housing

These areas are concerned with the planning, regulation, and management of towns and buildings and their surroundings

- *Examples of activities:* arrangement of new finance for construction and development, development of new building projects, town planning, country planning, building control, health and safety, estate management, housing management.

Built environment activities

The progress of a construction and built environment project involves the general stages and processes described below. These activities need people and the occupations in the built environment sector can also be placed against the processes. The activities are ongoing because the 'final' process of successfully using a building or other facility continually involves projects of maintenance and improvement. Even if no new buildings are being built in a town, the work in the built environment continues.

Activity stages

Proposals: client need, finance, location, scheme design, planning, approvals.

Preparation: detail design, costing, tenders, project planning.

Operations: site preparation, supplies, construction, installations, finishing.

Facilities and estate management: efficient use, maintenance, refurbishments, adaptation.

There are many occupations and careers available in construction and the built environment. These occupations can be grouped according to the activities in which they specialise. People can also be grouped according to the type of organisation which employs them, such as private industry or local government for example, or grouped by their qualification, such as architects or bricklayers.

Architecture

The various processes of architecture aim to design buildings that satisfy the client, that function properly, and that are pleasing to the eye for the whole life of the building.

Design teams are usually led by *registered architects* who also act as professional advisers to the client who has commissioned the building. Architects can work in private practice, local and national government, with property developers and within construction companies.

Architects are supported by other members of the design team, including qualified architectural technicians who have developed skills in producing drawings by computer-aided design (CAD) or hand draughting. An architect may be represented on site by a *clerk of works*.

Capabilities for architects
Visual awareness
Spatial awareness
Creativity
Social awareness
Technology knowledge
Drawing skills
Organisational skills
Communication skills

Engineering

Capabilities for engineers
Technology knowledge
Mathematical skills
Design skills
Drawing skills
Problem-solving skills
Organisational skills
Communication skills

Civil engineers specialise in the design and construction of engineering projects such as roads and bridges. Groups of engineers tend to specialise in areas of work such as ground and foundation (geotechnical) engineering, environmental engineering, railways, highways, dams and irrigation.

Structural engineers develop a stable structural design for the building or other structure so that each element can safely withstand the loads and stresses which are placed upon it. The structural engineer works closely with the architect to produce a design which is then checked and approved by the local authority.

Building services engineers are engineers who design and manage the installation of services such as electricity, water and fuel supplies, lifts, security systems, lighting, heating, air conditioning and refrigeration. Building services engineers are employed as consultants, by national bodies, by building contractors, by suppliers of equipment, and by the client.

Surveying

Capabilities for surveyors
Legal and contract
    knowledge
Planning knowledge
Technology knowledge
Estimating abilities
Site measurement abilities
Organisational skills
Communication skills

Building surveyors work with existing buildings and supervise the work of maintaining and repairing buildings or adapting buildings for new uses. They work with architects and civil/structural engineers to draw up plans and to prepare budgets for clients.

A *quantity surveyor* is an adviser on the costs of construction activities, the forms of contract used for projects, the value of work done, and the methods of payment for projects. The quantity surveyor has wide knowledge of legal and contract procedures and supervises the measurement and estimating techniques used for costing construction work.

Land surveyors collect and manage data about the natural and the built features of the land. The information may be recorded and presented in the form of maps or used for making decisions about construction projects.

Surveyors have the opportunity to work in private practice or in multidisciplinary firms which also employ architects and engineers. Other opportunities include construction companies, the public sector, and industrial or commercial organisations with property holdings.

Construction management

The main 'contractor' or principal contractor for a construction project agrees, by a legal contract, to carry out the work

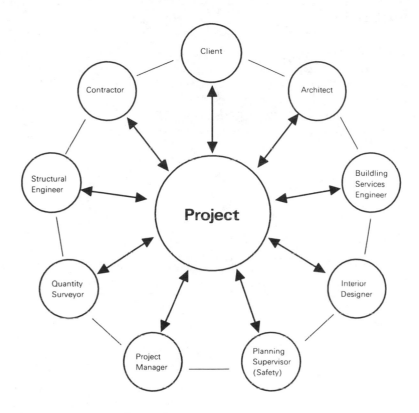

Figure 3.1   Construction project team.

required by the client who is paying for the work. The main contractor may choose to pay *subcontractors* to do some parts of the project. A large contracting firm employs a wide range of people, some of whom are described below.

Before a project starts work on the ground the main office of the contractor may have been involved in the design details, and will employ *quantity surveyors* and *estimators* to measure the work, calculate expected costs, tender a price and successfully gain the contract. *Buyers* will have to find convenient sources of materials and transport, and negotiate prices and delivery dates.

The *contracts manager* is responsible for the overall running of one or more contracts and acts as the link between the main office and the *project managers* on site. The s*ite agent* or *site manager* has responsibility for all personnel on site and their activities. Typical activities include planning work, labour relations, liaisons with architects and engineers, and maintaining links between the site and the contractor's main office.

Large construction projects will also have a *site engineer* who has responsibility for 'setting out' the exact positions of all buildings, roads and drains. *Surveyors* working on site have

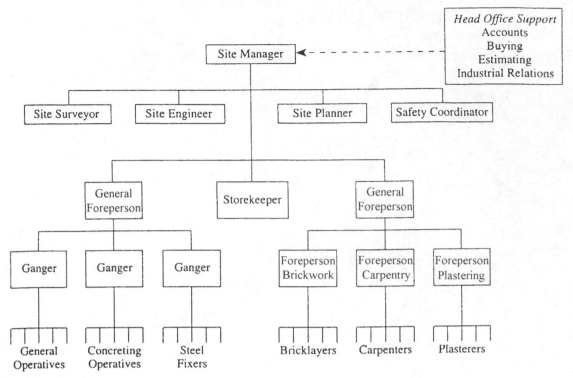

Figure 3.2 Typical site organisation structure.

Craft activities

Carpentry
Joinery

Bricklaying
Plastering
Roof and floor tiling

Formworking
Scaffolding
Roofing
Mastic asphalt work
Roadwork

Plastering
Roof, wall and floor tiling
Painting and decorating
Glazing
Shopfitting

Plumbing
Electrical listallation
Heating and ventilating
    installation
Refrigeration and
    air-conditioning

responsibility for measuring the amount of work completed at regular stages, such as each month, so that stage payments can be claimed by the contractor from the client. The contractor can then pay for supplies of materials and for subcontractors.

Skilled crafts

Craftspeople have specialist skills to perform certain tasks and to work with specific materials in a given 'trade', such as carpentry.

The *trades supervisor* is in charge of a particular trade, such as bricklaying, and has responsibility for keeping all craftspeople operating at maximum efficiency.

The trades listed in the margin are involved in one or more stages of construction operations. Not all projects will involve all types of trades and some people may be qualified in more than one trade.

Industry organisations

Institutions and societies

Institutions and societies are groups formed within a particular professional or occupation. Their aims and activities may include the following:

- Regulation of qualifications
- Promotion of education and standards
- Assurance of members' work.

Some institutions and societies

Architects and Surveyors Institute
Association of Building Engineers
Association of Consulting Engineers
Board of Incorporated Engineers & Technicians
British Institute of Architectural Technologists
British Institute of Facilities Management
The Chartered Institute of Arbitors
The Chartered Institute of Building
Chartered Institute of Building Services Engineers
Chartered Institute of Housing
Chartered Institute of Transport
Chartered Institution of Water & Environmental
    Management
Federation of Master Builders
Guild of Bricklayers
Guild of Incorporated Surveyors
Guild of Master Crafts
Incorporated Society of Valuers & Auctioneers
Institute of Asphalt Technology
Institute of Building Control
Institute of Carpenters
Institute of Clerk of Works
Institute of Concrete Technology
Institute of Construction Management
Institute of Domestic Heating & Environmental Engineers
Institute of Highway Incorporated Engineers
Institute of Maintenance & Building Management
Institute of Plumbing
Institute of Project Management
Institute of Refrigeration
Institute of Roofing

(continued)

Some institutions and societies (*continued*)

Institute of Shopfitting
Institution of Civil Engineering Surveyors
Institution of Civil Engineers
Institution of Electronics & Electrical Incorporated
    Engineers
Institution of Highways & Transportation
Institution of Structural Engineers
Landscape Institute
Royal Engineers
The Royal Incorporation of Architects in Scotland
Royal Institute of British Architects
Royal Institution of Chartered Surveyors
Royal Town Planning Institute
Society of Surveying Technicians
Society of Town Planning Technicians
The Worshipful Company of Painters and Stainers

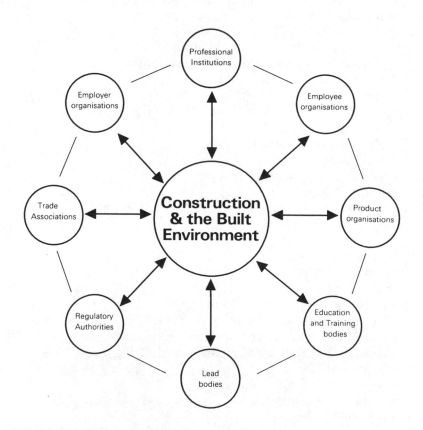

Figure 3.3   Industry organisations.

Employer and employee organisations

Organisations for employers are formed by groups of firms in the construction sector who have similar interests. Their activities typically include:

- Promoting the interests of employers
- Informing Government of employer needs
- Coordinating the interests of employers.

Organisations for employees are formed by groups of employees working in construction and its associated activities. Their activities typically include:

- Promoting the interests of employees
- Informing Government of employee needs
- Coordinating the interests of employees
- Negotiating with employers about employee interests.

Some employer and employee organisations

Employers
British Plumbing Employers' Council
Building Employers' Confederation
Federation of Civil Engineering Contractors
Heating & Ventilating Contractors' Association
Joint Contractors' Tribunal
Local Government Management Board
National Federation of Building Trades Employers
National Federation of Demolition Contractors
National Specialist Contractors' Council
Scottish Building Employers' Federation
Thermal Insulation Contractors' Association

Employees
Amalgamated Engineering Union
Electrical, Electronic, Telecommunications & Plumbing Union
National & Local Government Officers Association
Transport & General Workers Union

Industry, training and trade bodies

Some groupings in the fields of construction and the built environment include those with wide aims such as the ones outlined below:

- Informing Government of industry needs
- Coordination of relationships within the industry
- Promotion of the interests of the construction industry.

Industry Training Organisations (ITOs) have a prominent role among industry bodies to promote and administer schemes for funding training. Overall bodies, such as the Construction Industry Standing Conference (CISC), work with industry in the development and implementation of occupational standards leading to recognised awards such as NVQs.

Good technical practice in the industry is promoted by research organisations and by bodies which have the authority to certify that products perform to certain standards; of fire resistance for example. Trade associations between producers of materials or components are organisations whose activities include:

- Development and promotion of products
- Promotion of quality and of standards
- Coordination of interests.

Some industry, training and trade bodies

Coordination and training
Architects Registration Council of the UK
British Association of Construction Heads
Conference on Training in Architectural Conservation
Construction Industry Council
Construction Industry Standing Conference
Construction Industry Training Board
Electrical Installation Engineering Industry Training
  Organisation
Joint Contracts Tribunal
Joint Industry Board for Plumbing Mechanical
  Engineers Services
Quarry Products Training Council

Research and Products
Brick Development Association
British Board of Agrément
Building Research Establishment

(continued)

Some industry, training and trade bodies (*continued*)

Cement and Concrete Association
Centre for Window & Cladding Technology
Fire Research Institute
Timber Research and Development Association

Training and education

The industries of construction and associated activities in the built environment are a major source of employment in any country. The use of high technology in materials, construction techniques and services installations requires everyone involved to be well-trained and educated. For every person seen on a large construction site there are at least two others who have been involved in planning, design, preparation and management for those site activities. The previous sections of this chapter have outlined the roles of these people.

Governments also appreciate that the wealth of a modern country depends on citizens who are educated, well-trained for jobs now, and able to adapt to new roles in the future. The United Kingdom, for example, has national targets for minimum standards of education and training.

You should also have a personal target of education and training which will suit you for occupations now and in the future. If you are studying for a qualification you should be able to find yourself on one of the pathways and at one of the levels of education and training framework illustrated in Figure 3.4. The information is focused on the framework for the UK but similar schemes apply to most countries.

Construction is an international industry and it is common for people working in the construction industry to move around the world and work in other countries. At the moment it is not always easy to have your UK qualification recognised in another country but there are moves towards common recognition of standards and qualifications among countries.

The awards for education and training can be grouped in the following pathways:

- Occupational awards
- Vocational awards
- Academic awards.

All pathways offer the chance of improving your qualifications and prospects, and there are many links between pathways which let you move between areas according to the needs of

Construction and the Built Environment

PROGRESSION ROUTES

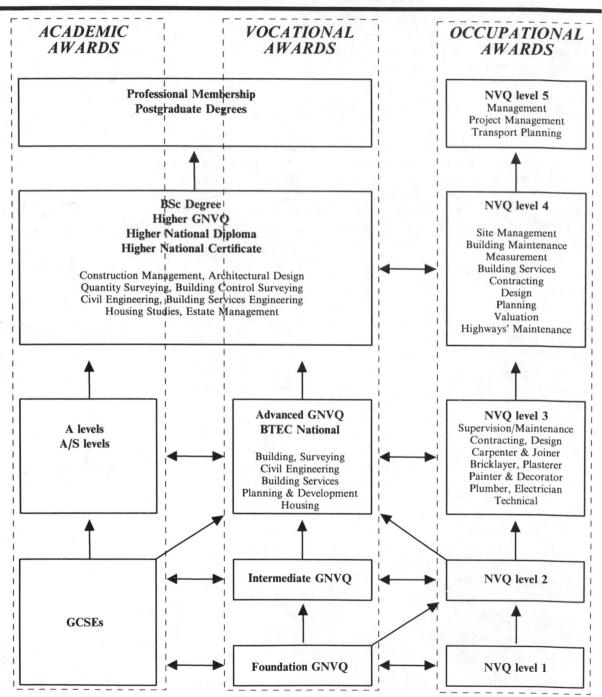

Figure 3.4 Progression pathways.

your job or with changes in your interests. Even if you decide not to work in construction and the built environment you will find your qualifications always have a good 'transfer' value. In other words, the skills and knowledge that you learn from employment in construction can also be used in other work areas.

Occupational awards

Occupational awards focus on the performance and knowledge needed to carry out particular tasks in a job. Competencies in tasks can usually be assessed in the workplace. The gaining of an award indicates that a person is qualified in an area of skill and may be termed a qualified craftsman, craftswoman, operator, technician, supervisor or other title.

- *Typical occupational areas*: craft skills, office skills, supervision skills
- *Typical occupational awards*: NVQs, City & Guilds, RSA.

Vocational awards

Vocational awards involve both knowledge of a subject and how that knowledge is used in practice in vocational areas such as engineering, town planning, or building management. Vocational qualifications give a good experience of work in your chosen area by means of links between the course and employers and, often, by sponsored study.

The award indicates a qualified technical officer or junior manager and the practical experience helps progression to professional qualifications.

- *Typical vocational areas*: building studies, civil engineering studies, building services engineering studies, surveying studies, urban development studies, estate management, housing studies
- *Typical vocational awards*: GNVQs, Applied A-Levels, Edexcel/BTEC National Certificates & Diplomas, Higher National Certificates and Diplomas.

Academic awards

Academic awards focus on areas of knowledge which can be learned at a school, college or university and do not yet involve significant practical experience. The awards can be used for progression to professional qualifications after suitable practical experience has been gained.

- *Typical academic areas*: science, mathematics, languages, architecture, engineering, building studies, surveying studies, urban development
- *Typical academic awards*: GCSEs, A/AS levels, degrees.

Professional qualifications

Some professional titles
Associate member
Chartered member
Fellow

A professional qualification recognises both high levels of knowledge and practical experience in a chosen area. Membership of a professional body may be legally necessary for some activities and often provides insurance cover.

- *Typical professional awards*: registered architect, chartered builder, chartered engineer, chartered surveyor, chartered town planner, chartered housing practitioner.

Updating

Training and education never stops! Even the most qualified person needs to learn about changing technology, improved materials and methods, and about changing regulations. Electrical and gas installation and structural engineering are areas where it is obviously dangerous unless people have up-to-date qualifications.

Everyone in the industry needs updated knowledge and the various organisations run courses at all levels for different sectors in construction and the built environment. Many professional bodies require their members to undertake a minimum number of hours of Continuing Professional Development (CPD) so that they can continue as reliable and credible practitioners.

Key words for Chapter 3, Organisations and Careers

The following is a list of some keywords used in this chapter. Use the list to test your knowledge and, if necessary, consult the text to learn about the terms.

Building services	Estate management	Structural Engineers
Building surveyor	ITOs	Trades supervisor
CPD	Professional	Vocational awards
Employer organisations	qualifications	

4 Environmental Science

This chapter studies the internal environment we create inside buildings and the scientific background to this environment. A general aim is to ensure that a building provides a pleasant space where humans can live and work in comfort, without wasting energy. Topic areas involved in the assessment and control of the internal environment include those listed below:

- Thermal insulation
- Heat losses and gains
- Energy use
- Thermal comfort conditions
- Air quality
- Sound comfort
- Noise measurement
- Lighting measurement
- Lighting conditions.

Thermal energy in buildings

Buildings are the single largest consumer of the total energy used by a country. Most of this energy is used for heating and lighting, although in some office buildings more energy is used for cooling on a summer's day than for heating on a winter's day.

Heat terms

Heat
Heat is a form of energy.

Unit: joule (J).

(continued)

Heat terms (*continued*)

Temperature
Temperature is the condition of a body that determines whether heat will flow from the body.

Unit: degree Kelvin (K) is the formal SI unit
and
degree Celsius (°C) is in common use.

One degree Celsius is the same size as one degree Kelvin but they are connected by the following relationship:

- temperature Kelvin = temperature Celsius + 273

Heat transfer
Heat will move from an area of higher to lower temperature by the following mechanisms:

- Conduction – in solids, liquids, gases
- Convection – in liquid, gases
- Radiation – in gases, vacuum (across space).

Thermal insulation

Thermal Insulation in buildings is the use of materials and techniques to prevent the transfer of heat by conduction, convection and radiation. For example, fibreglass wool and similar porous materials hold plenty of still air which is a poor conductor of heat. Draught-proofing works by preventing air moving (convection). Shiny surfaces, such as aluminium foil, work by making radiation difficult.

Good thermal insulation is a major factor in reducing heat loss from a building and modern building regulations require that insulation is considered at the design stage. Good thermal insulation also reduces the flow of heat *into* a building and helps prevent overheating.

U-values

Heat passes through any part of the external 'shell' of the building, such as a wall, by a combination of conduction, convection and radiation. These effects are combined into a single measurement called the *overall thermal transmission coefficient* or *U-value*.

Table 4.1 Typical *U*-values

Element	Composition	U-value ($W/m^2 K$)
Traditional solid wall	solid brickwork	2.3
Early cavity wall	brickwork, cavity, brickwork	1.5
Modern cavity wall	brickwork, cavity with insulation, lightweight concrete block	0.6 to 0.3
Modern timber frame wall	brickwork, cavity, insulation, airgap, plasterboard	0.45
Traditional pitched roof	without insulation	2.0
Modern roof	with insulation	0.25
Window	with single glazing	5.7
Modern window	with double glazing	2.8

*A **U-value** is a measure of the overall rate of heat transfer through a particular section of construction.*

Unit: $W/m^2 K$ or $W/m^2 °C$ (same units)

The following effects are included in a *U*-value:

- Conduction through each 'layer' of material
- Convection and radiation in any gap or 'cavity'
- Convection and radiation at the inside and outside surfaces of the construction.

U-values for a particular type of wall, roof or other form of construction can be calculated by using data for the above effects (see Table 4.1). The aim of this chapter is to be aware of typical *U*-values and to use them for heat loss calculations.

- **Lower *U*-values mean better thermal insulation**

Heat losses

Heat losses from a building occur by a number of mechanisms, as shown in Figure 4.1. An uninsulated house built earlier this century has a high total heat loss with most of the loss via the walls and roof. A modern building has a greatly reduced total heat loss

Sources of heat loss
Wall loss
Roof loss
Floor loss
Window loss
Draughts
Ventilation

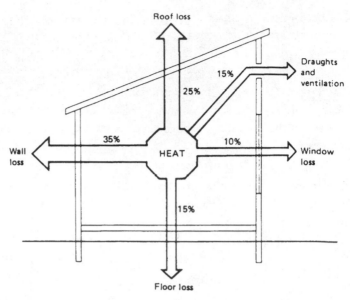

Figure 4.1   Types of heat loss from a building.

Heat loss factors
Insulation of building
External area of building
Temperature difference
Air change rate
Exposure to climate

with a high proportion of that reduced total heat loss happening via the ventilation system or when doors are opened and shut.

The amount of heat lost from a particular building depends on the heat loss factors listed in the margin. Good insulation, indicated by a low *U*-value, is a common target for reducing heat loss but these losses are further reduced by minimising the external area of the building, such as when a 'party' wall is shared between two buildings. Because outside temperatures depend on the weather, the temperature difference between inside and outside can only be reduced by the inhabitants reducing the temperature settings inside the building.

The air change rate can be reduced by sealing unwanted air gaps and reducing ventilation through windows. 'Exposure' to climatic effects increases when buildings rise in height or are built on high unprotected sites where wind currents increase the heat lost by convection.

It is convenient to consider the heat losses from buildings as two types of loss, fabric loss and ventilation loss, which can be calculated by using similar techniques.

Fabric heat loss

Fabric heat loss from a building is caused by the transmission of heat through the 'external envelope' made by the walls, roof, floor and windows.

For most purposes it is reasonable to assume that the temperatures inside and outside the building are steady or have an average value. The rate of heat lost from each part of the building can then be calculated using the following formula:

$$P = UA(t_i - t_o)$$

where

    P is heat loss/time or 'power loss' (W)
    U is U-value of that part of the building (W/m^2 °C)
    A is area of that element of the building (m^2)
    t_i is temperature of inside environment (°C)
    t_o is temperature of outside environment (°C).

Ventilation loss

Ventilation heat loss from a building is caused by the loss of warm air and its replacement with cooler air which needs to be heated. The cooler air may enter by 'infiltration' through gaps in construction, by the opening of doors, or by regular ventilation through an open window.

The rate of heat loss for ventilation heat losses is calculated by the following formula:

$$P = \frac{c_v N V (t_i - t_o)}{3600}$$

where

    P is heat loss/time or 'power loss' (W)
    c_v is volumetric specific heat capacity of air (common working
       value = 1300 J/m^3 K)
    V is volume of the room or space (m^3)
    t_i is temperature of inside air (°C)
    t_o is temperature of new air (°C).

Total heat energy losses

It is reasonable to use average temperatures for the calculation of daily heat losses. However, when calculating maximum heat losses, an external design temperature such as -1°C is suitable. This approach is used, for example, for choosing the appropriate capacity of heating equipment.

The total heat loss from a room or a complete building is built up by separately calculating losses for each part of the fabric across which there is a temperature difference. If there is no difference in temperature across a wall or other area then there is zero heat loss. The various fabric losses are best calculated in a table or computer spreadsheet, as shown in the example, and added to give a total fabric heat loss. The total heat loss figure is then added to the ventilation heat loss to give the total rate of heat loss.

The figures for rates of heat loss or 'power loss' are in terms of watts which is energy per second. To find the energy loss over a given period the following formula is used:

$$E = Pt$$

where

E is energy (J)
P is power loss (W)
t is time (seconds).

Energy is also expressed in megajoules (MJ) and sometimes in kilowatt hours (kWh):

1 MJ = 1 000 000 J
1 kWh = 3.6 MJ

Sample heat loss calculation

A simple building is 4 m long by 3 m wide by 2.5 m high. The walls contain one window of 2 m by 0.6 m and there is one door of 1.75 m by 0.8 m. The construction has the following U-values in W/m^2 K: walls 2.5, windows 5.6, door 2.0, roof 3.0, floor 1.5. The inside temperature is kept at 18°C while the outside temperature is 6°C. The volumetric specific heat capacity of air is taken to be 1300 J/m^3 °C. There are 1.5 air changes per hour. Calculate the total rate of heat loss for the building under the above conditions.

Calculate the areas and temperature difference for each part of the building

Make a table of the information and calculate the fabric power loss using the formula:

$$P = UA(t_i - t_o)$$

(continued)

Sample heat loss calculation (*continued*)

Element	U-value $(W/m^2 K)$	Area (m^2)	Temperature difference $(°C)$	Rate of heat loss (W)
Window	5.6	1.2	12	80.64
Door	2.0	1.4	12	33.6
Walls	2.5	35–2.6	12	972
Roof	3.0	12	12	432
Floor	1.5	12	12	216

Total rate of fabric heat loss   1734.24 W

Calculate the ventilation power loss using the formula:

$$P = \frac{c_v N V (t_i - t_o)}{3600}$$

With $c_v = 1300$, $N = 1.5$, $V = 4 \times 3 \times 2.5 = 30$, $t_i - t_o = 18 - 6 = 12$:

$$P = \frac{1300 \times 1.5 \times 30 \times 12}{3600}$$

So rate of ventilation heat loss = **195 W**.

Combine fabric losses and ventilation loss

Total rate of heat loss = fabric loss + ventilation loss

$$= 1734.24 \quad + 195$$

$$= \textbf{1929.24 W}$$

Heat gains

Even in winter a building gains heat at the same time that it is losing heat. These heat gains are from the Sun and from activities in the building.

Solar heat gains are affected by the position of the building, the season of the year and the local weather. Direct radiation through windows, when the Sun is lower in the sky, causes the largest solar gains for buildings in the British Isles.

Other heat gains are generated by activities and equipment in buildings that give off heat even though they are not intended for heating. For example, each person in a room emits over

Sources of heat gain
Solar heat through windows
Solar heat through walls
    and roof
Body heat from inhabitants
Heat from lighting
Cooking
Water heating
Electrical appliances
Machinery

Table 4.2   Typical heat gains

Type of source	Typical heat emission
For typical adult	
Seated at rest	90 W
Medium work	140 W
Lighting	
Fluorescent system (e.g. classroom)	20 W/m^2
Tungsten system (e.g. home)	40 W/m^2
Equipment	
Desktop computer	150 W
Photocopier	800 W
Gas cooker	350 W per burner
Colour TV	100 W

100 W of heat, so that 10 people provide more heat than a one-bar electric heater. Ordinary tungsten filament lamps or 'light bulbs' waste about 90 per cent of their electrical energy as heat.

Calculations of heat gains are relatively complicated but the use of standard figures (see Table 4.2) gives acceptable results.

Energy requirements

To maintain a constant temperature inside a building, the heat losses and gains need to be balanced by using extra heating or cooling. Both heating and cooling use energy. For most buildings in winter the energy which needs to be supplied by the heating plant is given by the following expression of *heat balance*:

$$Energy\ needed = Heat\ losses - Heat\ gains$$

Table 4.3   Typical heating system efficiencies

Type of system	House efficiency
Central heating (gas, oil)	60–70 per cent
Gas radiant heater	50–60 per cent
Gas convector heater	60–70 per cent
Electric fire	100 per cent

Note: The efficiency of an electrical heating appliance is 100 per cent *within* the building, but the overall efficiency of the generation and distribution system is about 35 per cent.

The heat energy needed for buildings is commonly obtained from fuels such as coal, gas and oil, even if this energy is delivered in the form of electricity. The amount of heat finally obtained in the building depends on the energy content of the fuel and the efficiency of the system which converts and distributes the heat (see Table 4.3).

Comfort conditions

Buildings such as houses and offices need to be comfortable places where we can live and work. This section gives an introduction to the conditions and measurements which help to define human comfort. These conditions are a starting point for the study of the building engineering services that help to maintain comfort.

Comfort factors
Thermal
Air quality
Sound
Lighting

Thermal comfort

The thermal comfort of humans depends on factors which vary from person to person. The heat produced by food energy in the body has to be lost at the best rate to keep the body at constant temperature. The amount of heat produced by a person depends on personal variables of body chemistry, body size, age, gender, type of activity and clothing.

The rate at which the body exchanges heat with its surroundings and feels comfortable depends on the following factors:

Thermal comfort aims
Correct temperatures
Minimum variations
Correct humidity
Low air movement

Thermal comfort factors
Activity
Clothing
Age
Male/female
Temperatures
Air movement
Humidity

- Temperature of the air
- Temperature of surrounding surfaces
- Air movement
- Moisture in the air.

Air temperature (see Table 4.4) is measured by a dry bulb thermometer, which might also be called a 'plain' thermometer. The bulb of the thermometer needs to be protected from air movement and from radiant effects caused by the Sun or by body heat.

The surfaces surrounding a person produce radiant effects on the skin and can cause discomfort if they are significantly different from the air temperature. Cold surfaces of windows, for example, can cause discomfort even though the air temperature is satisfactory. Some parts of the body, such as the ankle and the

Table 4.4   Thermal comfort measurements

Measurement	Some comfort values	Significant values	Instrument
Air temperature	21°C living room 20°C offices 18°C shops	0°C – freezing 100°C – boiling	Dry bulb thermometer
Relative humidity	40 to 70% RH	0% – dry 100% – saturated	Hygrometer

neck, are also sensitive to slight movements of air which can be caused by natural convection currents in a room.

Moisture in the air can be specified in various ways but relative humidity relates well to human comfort (see Table 4.4). Humidity is measured by instruments called *hygrometers* which may use electronic probes or may use a combination of wet and dry bulb thermometers.

Air quality

Ventilation aims
Oxygen supply
Carbon dioxide removal
Odour removal
Pollutant removal
Heat removal

Ventilation systems
Natural ventilation
Mechanical ventilation
Air conditioning

As humans we need certain qualities of air for preserving life and for comfort. Breathing requires a minimum supply of oxygen and, more importantly, the removal of carbon dioxide. However, long before there is any danger to life we will object to uncomfortable temperatures, odours and contaminants in the air from activities like smoking, cooking and washing.

Ventilation systems maintain air quality by removing 'stale' air and replacing it with 'fresh' air. Open windows are a common form of natural ventilation which allow convection currents to change the air. Mechanical ventilation provides more control by using fans to supply new air or to exhaust old air. Air conditioning has extra equipment which allows the temperature and humidity of the air to be adjusted for comfort conditions. All ventilation systems involve loss of heat but mechanical systems and air conditioning give opportunities to recycle heat from exhausted air.

Rates of ventilation may be specified in several ways: in terms of the entire room such as air changes per hour; or in terms of individuals such as litres per second per person (see Table 4.5). Prediction of ventilation rates can be made by knowing the capacities of fans and ducts, or by reference to tables of typical rates. If necessary, moving air can be measured by an *anemometer*, similar to that used for wind measurements.

Table 4.5 Table of ventilation measurements

Measurement	Some comfort values	Significant values	Prediction
Air changes per hour (ach)	0.5 ach – bedroom 1 ach – living room 3 ach – WC	0 ach – airtight	Published values and calculation
Air supply rate	8 litres/s per person – minimum	50 l/s/p – smoking 60 l/s/p – kitchen	Published values and calculation

Sound comfort

Comfortable surroundings of sound and noise, which may also be called *aural* comfort, are affected by the habits and preferences of individual people, rather like thermal comfort. There are also

Aural environment aims
No dangerous sound levels
Minimum background level
Suitable room acoustics

Decibels	Typical examples
140	Threshold of pain
130	
120	Threshold of discomfort
110	
100	Pneumatic drill
90	
80	Busy traffic
70	
60	Conversation
50	
40	Living room
30	
20	Quiet countryside
10	
0	Threshold of hearing

Figure 4.2 Decibel scale of sound levels.

certain levels of sound which cause damage to everyone's hearing system, even if this damage is not recognised at the time that it occurs.

Environmentally, noise is defined as *unwanted* sound. This definition means that good music can be defined as noise if it is played at levels or at times that other people consider unreasonable.

Problems from noise
Hearing loss
Lowered quality of life
Communication difficulties
Distraction from tasks
Expense of counter measures

Sound terms

Sound
Sound is a form of vibrational energy which can pass through air, liquids and solids.

Frequency
Frequency is the rate at which sound waves vibrate.

Unit: hertz (Hz) which is vibration cycles per second.

Humans interpret lower frequencies as *bass* notes and higher frequencies as *treble* notes.

Typical frequencies
    20 Hz lower hearing limit
    27 Hz lowest note on piano
   262 Hz 'middle C' in music
  1000 Hz radio time signal
  4186 Hz highest notes on piano
  7000 Hz limit of AM radio (variable)
17 000 Hz upper hearing limit (variable)

Sound levels
Sound levels are measurements of sound strengths on a scale which matches the response of human hearing.

Unit: decibel (dB) – a scale based on logarithmic ratios.

A measurement of dB(A) indicates that sound of mixed frequencies was measured using a standard 'A-scale' setting on a sound level meter.

The sound effects produced by various sound level changes are given in Table 4.6.

Table 4.6 Table of sound level changes

Sound level change	Sound effect heard
±1 dB	negligible
±3 dB	just noticeable
+10 dB	twice as loud
−10 dB	half as loud
+20 dB	four times as loud
−20 dB	one quarter as loud

Noise measurement

In addition to individual sensitivities of hearing and habit, the following factors need to be considered when assessing noise:

- *Surroundings*: sleeping, talking, hammering, for example have different requirements
- *Frequency content*: some frequencies, such as high whines, are more annoying than others
- *Time duration*: a short period of high noise may be less annoying than long periods.

Various systems of noise measurement have evolved for different types of situation (see Table 4.7). Noise indices start with a measurement of sound level in decibels and then include factors of frequency content and time duration.

Table 4.7 Table of noise measurements

Measurement	Unit	Some significant values	Instruments
Sound level	dB(A)	40 dB(A) – quiet living room 60 dB(A) – conversation 80 dB(A) – busy high street	Sound level meter
Equivalent continuous sound level	L_{eq}	L_{eq} (8 hour) = 75 dB(A) is one limit for construction site noise	Sound level meter with statistical ability
Occupational noise index	$L_{EP,d}$	$L_{EP,d}$ = 85 dB(A) is one limit for industrial exposure over 8 hours	Sound level meter with statistical ability
Traffic noise level	L_{10}	$L_{10,18hr}$ = 68 db(A) is one limit for roadside noise	Sound level meter with statistical ability

Table 4.8   Reverberation times

Purpose of room	Suitable reverberation time	Measurement
Speech	0.5 to 1 second	Measured by timing sharp sounds, or predicted by formula
Music	1 to 2 seconds (Larger volumes need longer times)	

Acoustic quality

Acoustic aims
Adequate sound levels
Even distribution
Suitable reverberation time
Minimum external noise
Absence of unsuitable
    reflections

Acoustics is concerned with sound *within* a room or enclosed space and deals with qualities such as evenness, clarity and fullness. Acoustic studies are less concerned with the sound insulation of a room, unless sound from outside the room is at an annoying level.

Reverberation is the continuing presence of an audible sound after the source of sound has stopped.

Reverberation can be assessed by measuring the *reverberation time*. The reverberation time in a room depends on the absorption of the surfaces and the distances between them. Shorter reverberation times are needed for clarity of speech while longer reverberation times are considered to enhance the quality of music.

Noisy rooms, echoing corridors and clattering restaurants can be avoided by a simple knowledge of acoustics. For good acoustics in medium to large-sized rooms the following features are helpful:

- Irregular surfaces or decorations help to spread sounds evenly

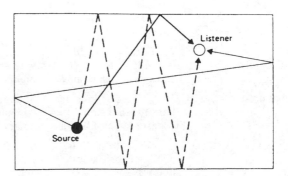

Figure 4.3   Reverberation in room.

- Reflections by surfaces close to the source help to reinforce the sound
- Absorption by porous surfaces and by flexible panels, such as ceilings and windows, reduce the reverberation time
- Absorption by surfaces far from the source help to prevent unwanted long reflections.

Lighting

The effects of light, from both artificial and natural sources, play a major role in our use and appreciation of the built environment. Artificial lighting in buildings also accounts for a significant proportion of energy use in buildings, and the prediction and measurement of lighting is therefore an important aspect of building design.

The quantity of light on a surface is usually the main design consideration but the other factors include colour, glare and modelling qualities. For example, some very efficient street lamps would give enough light in a home but the other light qualities of those lamps are unacceptable in the home.

Lighting purposes
For tasks
For moving about
For display
For security

Lighting methods
Artificial lighting: by lamps
Natural lighting: by windows

Lighting criteria
Types of source
Quantity of light
Colour effects
Glare effects
Directional modelling
Energy use
Costs

Lighting terms

Light
Light is energy in the form of electromagnetic radiation which can be detected by the human sense of sight.

Luminous intensity or source strength, is measured in *candelas* (cd).

Luminous flux or light flow is measured in *lumens* (lm).

Illuminance or light concentration on a surface is measured in *lux* (lx).

Colour
Light contains a range of wavelengths which the eye interprets as a *spectrum* of different colours. There is a continuous spread of hues but these are often described as the seven colours of the rainbow:

Red   Orange   Yellow   Green   Blue   Indigo   Violet

Lighting measurements

The primary consideration in lighting design is the quantity of light on a certain surface, such as a table or floor. The concentration of light is specified by *illuminance* and is measured in units of *lux*.

The illuminance needed for a particular task depends on the visual difficulty of the task and the type of performance expected (see Table 4.9). The speed and accuracy of various types of work are affected by the level of illuminance supplied. Values of standard service illuminance levels are recommended for a variety of interiors and tasks.

Daylight factor

For design purposes, natural light is assumed to come from an overcast sky and not from direct sunlight. The natural light through windows into a room is usually only a small fraction of the total light available from a complete sky.

The level of illuminance provided by the sky varies as the brightness of the sky varies so it is not possible to specify daylight as a fixed illuminance in lux. Instead, the daylight factor, in percentage, is used. The daylight factor is a ratio which remains constant even when the sky changes.

Table 4.9 Lighting measurements

Measurement	Some recommended values	Significant values	Instruments
Illuminance	50 to 150 lx on floors 500 lx on office desk 750 lx on drawing board 500 lx for assembly work 1000 lx for electronic work	0.1 lx – moonlight 5000 lx – overcast sky 50 000 lx – bright sunlight	Light meter (photometer)
Daylight factor	2% minimum on desk 5% average on desk	20% – inside near window 100% – outside unobstructed	Prediction from scale drawings, computer programs
Colour temperature	2700 K – tungsten filament lamp 3000 to 6000 K – fluorescent lamp	2000 K – candle 5000 to 8000 K – overcast sky 12 000 to 24 000 K – clear sky	Prediction from manufacturers' data

Summary of comfort measurements

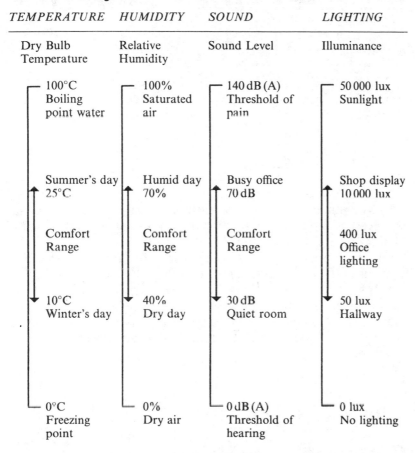

TEMPERATURE	HUMIDITY	SOUND	LIGHTING
Dry Bulb Temperature	Relative Humidity	Sound Level	Illuminance
100°C Boiling point water	100% Saturated air	140 dB (A) Threshold of pain	50 000 lux Sunlight
Summer's day 25°C	Humid day 70%	Busy office 70 dB	Shop display 10 000 lux
Comfort Range	Comfort Range	Comfort Range	400 lux Office lighting
10°C Winter's day	40% Dry day	30 dB Quiet room	50 lux Hallway
0°C Freezing point	0% Dry air	0 dB (A) Threshold of hearing	0 lux No lighting

Measurement ranges are not to scale.

Key words for Chapter 4, Environmental Science

The following is a list of some keywords used in this chapter. Use the list to test your knowledge and, if necessary, consult the text to learn about the terms.

daylight factor	hygrometer	occupational noise
degree Kelvin	illuminance	index
fabric heat loss	natural ventilation	reverberation
heat balance		*U*-value

5 Services Science

A supply of electricity is essential for creating and controlling the environment of a modern building. A supply of good water is an essential requirement for human life and is a fundamental service for a community and building.

The large-scale supply of both electricity and water requires decisions which can have major effects on the environment. For example, communities and landscapes are affected by lakes behind dams and by electricity supply lines. This chapter summarises the principles of science which are relevant to supplies of electricity and water, and looks at technical arrangements for their supplies.

Electricity principles

Conductors and insulators

If a material allows a significant flow of electric current then the material is termed a **conductor of electricity**. A material that passes relatively little current is termed an **insulator**.

Conductors

Solid conductors are materials whose free electrons readily produce a flow of *charge*. If the conductor is a liquid or a gas then the charge is usually transferred by the movements of *ions*. Common types of conductor are listed below:

- *Metals*: Examples include copper and aluminium cable conductors
- *Carbon*: Examples include sliding contacts in electric motors
- *Liquids and gases*: Current can flow when ions are present, such as in salty water or in a gas discharge lamp.

Insulators

Insulators are materials that have relatively few free electrons available to produce a flow of charge. Common types of insulator are listed below:

- *Rubber and plastic polymers*: Examples include PVC cable insulation
- *Mineral powder*: Examples include magnesium oxide cable insulation (MICC)
- *Oil*: Examples include underground cable insulation
- *Dry air*: Examples include overhead power line insulation
- *Porcelain and glass*: Examples include overhead power line insulation.

Electricity terms

Charge

Charge (*Q*) *is the basic quantity of electricity.*

Unit: coulomb (C).

Electric current

Electric current (*I*) *is the rate of flow of charge in a material.*

Unit: ampere (A).

Potential difference

For an electric current to flow in a conductor there must be a difference in charge between two points. This potential difference is similar to the pressure difference that must exist for water to flow in a pipe.

Potential difference (pd or *V*) *is a measure of the difference in charge between two points in a conductor.*

Unit: volt (V).

(*continued*)

Electricity terms (*continued*)

Resistance

$$Resistance\ (R) = \frac{Potential\ difference\ (V)}{Current\ (I)}$$

Unit: ohm (Ω) where $1\ \Omega = 1\ \text{V/A}$.

Note: This expression can be converted into two other forms as follows:

$$V = IR \quad \text{and} \quad R = \frac{V}{I}$$

Power and energy

The definitions for electrical energy and electrical power are the same as for other forms of power and energy, and they have the same units of joules and watts. Power is the rate of using energy and, because the volt is defined in terms of electric charge and energy, it is therefore possible to express power in terms of electric current (flow of charge) and voltage.

Electrical power

$$Power\ (P) = Current\ (I) \times Potential\ difference\ (V)$$

$$P = IV$$

Unit: watt (W)

where, by definition, 1 watt = 1 joule/second.

Two other useful expressions are obtained by combining the expressions $P = IV$ and $V = IR$:

$$P = I^2R \quad \text{and} \quad P = \frac{V^2}{R}$$

The power rating of an electrical appliance is often quoted in specifications and it is an indication of the relative energy consumption of the device. Some typical power ratings are given in the margin.

Typical power ratings

Electric kettle element	2500 W
Electric fire (1 bar)	1000 W
Colour television	100 W
Reading lamp	60 W
Calculator charger	5 W

Electrical energy

Power is defined by the energy used in a certain time, so it is also possible to express energy in terms of power and time.

$$Energy\ (E) = Power\ (P) \times Time\ (t)$$

$$E = Pt$$

Unit: joule (J)

where 1 joule = 1 watt × 1 second.
Note: The kilowatt hour (kWh) is an alternative unit of energy in common use for electrical purposes, where

$$1\,kWh = 1\,kilowatt \times 1\,hour = 3.6\,MJ$$

Worked example

A 3 kWh electric heater is connected to a 230 V supply and run continuously for 8 hours. Calculate: (a) the current flowing in the heater; and (b) the total energy used by the heater.

(a) Know $P = 3000\,W$, $I = ?$ $V = 230\,V$.

Using
$$P = IV$$
$$3000 = I \times 230$$

$$I = \frac{3000}{230} = 13.0$$

So current = 13.0 A

(b) Know $E = ?$ $P = 3\,kW = 3000\,W$ $t = 8\,h = 28\,800\,s$.

Using
$$E = Pt$$
$$E = 3000 \times 28\,800$$
$$= 86.4 \times 10^6\,J$$

(continued)

Worked example (*continued*)

Alternatively, using kilowatt hours:

$$E = 3 \times 8$$
$$= 24\,\text{kWh}$$

So energy $= 86.4\,\text{MJ}$ or $24\,\text{kWh}$

Electromagnetic induction

It is possible for an electrical effect to be induced in one circuit by the action of another circuit, even though there is no apparent contact between the two circuits. In fact there is always a very real link of magnetic flux.

Electromagnetic induction is the principle behind many important devices used for the generation, the transmission, and the application of electricity. Some examples are given below:

- *Generators*: for the production of electricity
- *Transformers*: for changing voltage
- *Induction motors*
- *Ignition coils*: to create a spark in car engines
- *Hi-Fi cartridges*: for playing records
- *Linkages* in electronic circuits.

General principle of induction
An electric current will be induced in a conductor which is subjected to a *changing* magnetic field.

This change in magnetic field may be caused by physical movement or by switching a circuit on and off.

Magnitude of induction
The size of an induced current depends on the following factors:

- Relative speed of movement of the magnetic field
- Strength of the magnetic field
- Length of conductor in the magnetic field
- Angle between the conductor and the field.

Alternating current electricity

A *generator* is a device that converts mechanical energy to electrical energy by means of electromagnetic induction. Most of the electricity used in everyday life is generated in this manner. The change in magnetic field necessary for induction is produced by moving a coil through a magnetic field, or by moving a magnetic field past a coil. Rotational motion is usually employed. Some types of generator may be known as a *dynamo* or an *alternator*.

Simple AC dynamo

The simplest type of generator is the AC dynamo, as shown in Figure 5.1.

The EMF and the current produced by this generator are therefore continuously 'alternating' with every revolution of the coil. The output of this 'alternator' is in the shape of a sine wave, as shown in the alternating current output of Figure 5.2.

Electromotive force (EMF)
A source of electrical energy
Unit: volt (V)

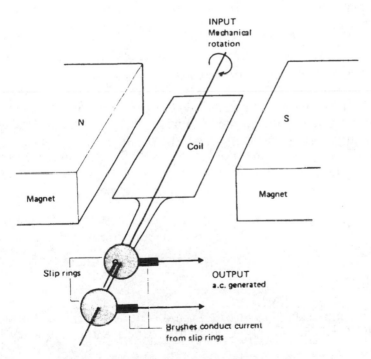

Figure 5.1 Simple AC dynamo.

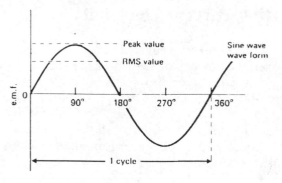

Figure 5.2   Alternating current output.

Practical AC generators

A large practical AC generator or alternator works on the same general principle as the simple AC generator but its construction is different.

The *stator* is the stationary frame to which are fixed the output windings that produce the current. No moving contacts are necessary to lead the current from the generator, so the construction is suitable for large supplies.

The *rotor* turns inside the stator and contains the magnetic field coils. The DC supply necessary for these electromagnets is supplied by a separate self-starting excitor dynamo, run on the same axle.

Alternating current properties

The EMF induced in an AC generator is constantly changing and reversing, and so the current produced by the EMF also changes to give the pattern shown in Figure 5.2. Some additional terms are needed to describe the nature of the alternating output.

Frequency
Frequency (f) is the number of repetitions, or cycles, of output per second.

Unit: hertz (Hz).

For public supplies in Britain and other countries in Europe the frequency is 50 Hz; in North America the frequency is 60 Hz.

(*continued*)

Alternating current properties (*continued*)

Peak value

The peak value is the maximum value of alternating voltage or current, measured in either direction. The peak values occur momentarily and only twice in a complete cycle, as shown in Figure 5.2.

RMS value

A **ROOT MEAN SQUARE** (*RMS*) *value of alternating current is that value of direc. current that has the same heating effect as the alternating current.*

A 1 kW fire, for example, produces the same heating effect using 240 V alternating current as it does us.   40 V direct current. The relationship between the values is iounu to be

RMS value = 0.707 Peak value

Transformers

A *transformer* is a device that uses the principle of electromagnetic induction for the following purposes:

- To step up or step down the voltage of an AC supply
- To isolate a circuit from an AC supply.

A major reason for using AC in electricity supplies is the relative simplicity and efficiency of transformers for changing voltage, as explained later in the sections dealing with power transmission. A transformer will not work on a DC supply. The construction of a simple type of transformer is shown in Figure 5.3.

Operation of transformer

The primary coil sets up a magnetic field which is continuously changing. The secondary coil experiences this changing magnetic field and produces an induced voltage (EMF) which can then be connected to a load. The induced voltage alternates with the same shape and frequency as the voltage of the supply frequency but the ratio of the two voltages is proportional to the ratio of the turns on the two windings.

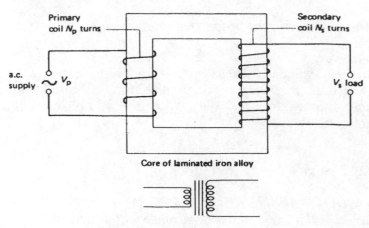

Figure 5.3   Step up transformer.

A transformer with twice the number of secondary turns as primary turns will, for example, *step up* the secondary voltage to twice the primary voltage. If the connections to such a transformer are reversed the transformer could be used as a *step-down transformer* which then halves the applied voltage. In an *isolating transformer* the primary and secondary voltage may be the same.

Transformer calculations

Transformer equation

$$\frac{V_s}{V_p} = \frac{N_s}{N_p}$$

where

V_s = EMF induced in the secondary coil (V)
V_p = EMF applied to the primary coil (V)
N_s = number of turns on the secondary coil
N_p = number of turns on the primary coil.

Efficiency equation
For the practical case where the transformer is less than 100 per cent efficient, the following formula is used:

$$I_s V_s = I_p V_p \times \text{efficiency factor}$$

(continued)

Transformer calculations (*continued*)

where

I_s = current in secondary coil (A)
V_s = EMF of secondary coil (V)
I_p = current in primary coil (A)
V_p = EMF of primary coil (V).

Transformers are efficient machines and their power output is almost equal to their power input. If, for simplicity, it is assumed that there are no power losses then the following relationship is true:

$$\text{Output power} = \text{Input power}$$

Worked example

A transformer has 600 turns on the primary coil and 30 turns on the secondary coil. An EMF of 240 V is applied to the primary coil and a current of 250 mA flows in the primary coil when the transformer is in use.
(a) Calculate the EMF of the secondary coil.
(b) Calculate the current flowing in the secondary coil. Assume that the transformer is 95 per cent efficient.

(a) Know $N_p = 600$, $V_p = 240$, $N_s = 30$, $V_s = ?$

Using
$$\frac{V_s}{V_p} = \frac{N_s}{N_p}$$

$$\frac{V_s}{240} = \frac{30}{600}$$

$$V_s = \frac{30}{600} \times 240 = 12 \text{ V}$$

So secondary EMF = **12 V**

(*continued*)

Worked example (*continued*)

(b)   Know $V_p = 240\,\text{V}$, $I_p = 250/1000 = 0.25\,\text{A}$, $V_s = 12\,\text{V}$, $I_s = ?$

Using $I_s V_s = I_p V_p \times 95/100$

$$I_s \times 12 = 0.25 \times 240 \times 95/100$$

$$I_s = \frac{0.25 \times 240}{12} \times \frac{95}{100} = 4.75\,\text{A}$$

So secondary current = **4.75 A**

Three-phase supply

The output from a simple AC generator, shown in Figure 5.2, is a *single-phase supply*. Practical supplies of electricity are usually generated and distributed as a three-phase supply, which is, in effect, three separate single-phase supplies each equally out of step. The generator is essentially composed of three induction coils instead of one, each coil being displaced from one another by 120 degrees.

   Three-phase supplies are economical in their use of conductors and can supply more power than single-phase supplies. Only three cables need to be used for the three-phase supply, instead of the six cables needed for three separate single-phase supplies. A device such as a motor which is designed to make use of the three phases receives more energy per second than those devices wired to a single phase; the three-phase motor is also smoother in its operation.

Power supplies

A public supply of electrical energy is one of the most important services in the built environment. Manufacturing, building services, transport and communications are all dependent on electricity supplies and any breakdown in the system greatly disrupts everyday life. Large electric power systems require significant investments in construction and they also have a considerable impact upon the appearance of the landscape.

A complete power system is a collection of equipment and cables capable of producing electrical energy and transferring it to the places where it can be used. A power system is made up from the following three main operations:

- **Generation**: the production of the electricity
- **Transmission**: the transfer of electrical energy over sizeable distances
- **Distribution**: the connection of individual consumers and the sale of electricity.

The overall energy efficiency of a power system is about 30 per cent.

Power stations

All power stations generate electrical energy by using electro-magnetic induction where an EMF is produced in a coil which experiences a changing magnetic field. The mechanical energy required is obtained using the heat from burning fossil fuels and from nuclear reactions, or obtained from the energy of moving water. A turbine is a device that produces rotational motion from the steam or running water and turns the axle of the generator.

A typical generator in a thermal power station turns at 3000 revolutions per minute and has an excitor dynamo mounted on the same shaft. The output varies but a large 500 MW generator set commonly generates 23 000 A at 22 kV. The cooling of the generators, by liquid or gas, is an important part of their design.

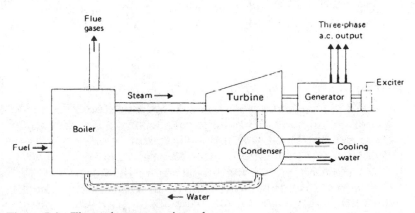

Figure 5.4 Thermal power station scheme.

Types of power station

Thermal power stations
Thermal power stations use heat energy to drive the generators. The heat is obtained by burning fuels such as oil, coal, or gas. The components of a thermal station are outlined in Figure 5.4. The boiler burns the fuel and heats water to produce high-pressure steam at high temperature. The steam is directed on to the blades of a high-speed turbine which produces mechanical energy and turns the generator. The steam is condensed and the water is returned to the boiler.

Nuclear power stations
Nuclear power stations are thermal stations where the heat energy is released from a nuclear reaction rather than by burning a fossil fuel. Radioactive elements, such as uranium and plutonium, have unstable nuclei which emit neutrons. These neutrons split neighbouring atoms, thus releasing other neutrons and producing heat. This fission reaction is controlled and prevented from becoming a chain reaction by using moderator materials, such as carbon, which absorb neutrons without reaction.

Hydroelectric power stations
Hydroelectric power stations use the kinetic and potential energy of running water to drive the generators. A large quantity of water at a height is required to provide enough energy. The original source of this energy is sunshine which lifts the water by means of evaporation. A dam usually provides both the head of water and a reservoir of stored water. The water flows down penstock pipes or tunnels and imparts energy to the water turbines which turn the generators.

Other types of electricity generation
Combined heat and power, CHP: using small power stations within communities. Buildings can be heated by the steam leaving the turbines, which is otherwise wasted.
Wind power: using the energy of natural wind.
Tidal or wave power: using the energy of the ocean.
Geothermal: using underground water and steam heated by the earth.

Transmission systems

Electrical energy has the useful property of being easily transferred from one place to another. A transmission system, as shown in Figure 5.5, consists of conducting cables and lines, stations for changing voltages and for switching, and a method of control. The energy losses in the system must be kept to a minimum.

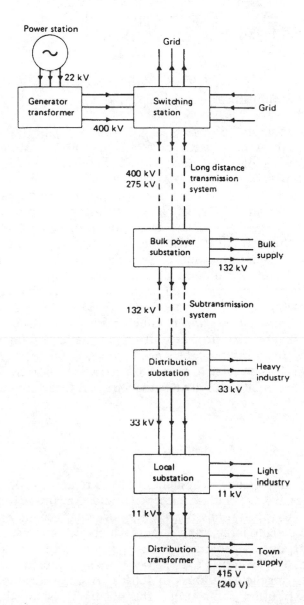

Figure 5.5 Typical power distribution.

Alternating current is used in nearly all modern power transmission systems because it is easy to change from one voltage to another, and the generators and motors involved in AC are simpler to construct than for DC. To obtain high transmission efficiency the current needs to be kept low because the heating losses in a line increase with the square of the current ($P = I^2 R$). Large currents also require thick conductors which are expensive and heavy. To transmit large amounts of power at low current there must be a high voltage ($P = IV$). Transformers are used for obtaining the necessary high voltages and wide air gaps are used to supply the high insulation that is needed at high voltage.

Transmission lines

The conductors for overhead transmission lines are made of aluminium with a steel core added to give strength. The transmission of three-phase supply requires three conductors, or multiples of three, such as six conductors.

Transmission towers or 'power pylons' are needed to keep the lines spaced in the air. Air is a convenient cheap insulator but higher voltages require larger airgaps in order to prevent short circuits through the air. The lines are suspended from the towers by solid insulators made of porcelain or glass. Transmission lines operate at voltages of 132 kV and above. The United Kingdom supergrid is at 400 kV and some countries have 735 kV systems.

In a buried cable the conductor must be insulated for its whole length and protected from mechanical damage. High-voltage underground cables need special cooling in order to be efficient. It is possible to bury 400 kV transmission cables but the cost is 10 to 20 times greater than for overhead lines.

Sub-stations and switching

Power transmission systems need provision for changing voltages, for re-routing electricity and for protecting against faults. At sub-stations the voltages are changed up or down as required using large transformers which are immersed in oil for the purpose of cooling as well as for insulation. At a power station, for example, a 1000 MVA transformer might step up 22 kV from the generators to 400 kV for the transmission lines. At distribution sub-stations the 400 kV is stepped down to 132 kV, 33 kV and 11 kV.

Transmission control

An electric power *grid* is a large system of interconnected power stations and consumers. The output from any power station is not dedicated to just one area but can be distributed to other areas as required.

This system allows the use of large power stations with lower operating costs to be used by all, especially at times of low demand such as at night. But one effect of large remote power stations is that surplus heat from the process cannot easily be used.

District distribution

At suitable junctions or at the ends of transmission lines the voltage is stepped down, as indicated in Figure 5.5. In the British system, the voltage is reduced from 400 kV (or 275 kV) to 132 kV. The electricity is distributed at this voltage by a sub-transmission system of overhead lines to the distribution sub-stations. At these stations the voltage is reduced to 33 kV and 11 kV for distribution by underground cable.

Large industrial consumers are supplied at 33 kV while smaller industrial consumers receive 11 kV. Small transformer stations in residential and commercial areas step the voltage down to the final 415 V three-phase, 230 V single-phase supply, which was explained in an earlier section of this chapter.

The three-phase supply is distributed by three-phase cables (red, yellow and blue) plus a neutral cable. Some commercial consumers are connected to all three cables of the supply. Households are connected to one of the phase cables and the neutral. Consumers are balanced between the three phases as evenly as possible by connecting consecutive houses to different phases in turn, for example.

Because perfect balance is not achieved the neutral cable carries a small amount of return current and is earthed at the distribution transformer. To ensure true earth potential, each consumer is supplied with an extra earth cable for attaching to the metal casing of electrical appliances. If an insulation fault causes a connection to this earth then a large current flows, immediately trips the fuse, and protects the user.

Water supplies

Water is essential for life on Earth and is also necessary for most of our agricultural and industrial activities. A supply of good water is therefore a fundamental service for a community and this supply requires major financial and engineering investments in the systems needed for the collection, storage, treatment and distribution of the water.

Principles of fluids

A *fluid* is a material whose particles are free to move their positions. Liquids and gases are both fluids and share common properties as fluids. The examples in this section are about understanding water but it is useful to remember that the same principles apply to the flow of air in ventilation ducts.

Pressure

An area which is submerged in a fluid, such as the base of a tank of water, experiences a pressure caused by the weight (which is a force) of water acting on the area of the base

*The **pressure** on any surface is defined as the force acting at right angles on that surface divided by the area of the surface.*

Pressure terms

Pressure formula
The pressure at any point in a fluid is given by the following formula:

$$p = \rho g h$$

where p = pressure at a point in a fluid (Pa)
ρ = density of the fluid (kg/m^3)
g = gravitational acceleration = 9.81 m/s^2
h = vertical depth from surface of fluid to the point (m).

Units of pressure
The pascal (Pa) is the SI unit of pressure where by definition, 1 pascal = 1 newton/square metre (N/m^2).

Because the values of density and gravitational acceleration in the pressure equation are usually constant, it is a common working practice to quote only the height or 'pressure head' associated with the pressure.

(continued)

Pressure terms (*continued*)

Principles of fluid pressure

- Pressure at any given depth is equal in all directions
- Pressure always acts at right angles to the containing surfaces
- Pressure is the same at points of equal depth, irrespective of volume or shape.

Energy of liquids

If a particle of liquid moves faster it gains kinetic energy. To compensate it must lose some of its pressure energy or potential energy. This is a particular example of the *general law of conservation of energy*, which states that the total energy of a closed system must remain constant. When this principle is applied to moving fluids it is stated as Bernoulli's theorem:

Bernoulli's theorem: *The total energy possessed by the particles of a moving fluid is constant.*

If the velocity of flowing water or air increases the kinetic energy must also increase. To keep the total energy constant the pressure energy must therefore decrease and there is a general rule for a moving fluid:

- Increase in velocity gives a decrease in pressure.

This principle, derived from Bernoulli's theorem, may initially seem surprising but it explains a number of important effects associated with moving liquids and gases. The shape of an aerofoil, such as an aircraft wing, causes the flowing air to have a higher velocity at the top of the wing than at the bottom. The increase in velocity produces a lower pressure on top of the wing, so that there is an upwards force on the wing.

Natural waters

The world possesses a fixed amount of water, which is found in various natural forms, such as oceans, lakes, rivers, underground waters, ice caps, glaciers and rain. This water plays an important part in maintaining the balance of the world's weather, especially through the presence of water vapour in

The water cycle
Evaporation:
    from the liquid state to
    water vapour
Condensation:
    liquid water appears as
    clouds of droplets
Precipitation:
    water droplets fall as rain,
    snow or hail
Run off:
    surface water and ground
    water

the atmosphere. Water is also essential for the growth of vegetation such as trees and food crops.

Humans need a small amount of essential drinking water but much greater amounts are used for washing and waste disposal, in homes, industry and commerce. The daily consumption of water in some cities is over 300 litres per person on average. The total supply of natural water in the Earth is enormous and should be adequate for our needs. However, local shortages do occur, especially when droughts are combined with poor management of resources. This section looks at the sources of water which are used for community water supplies and describes the qualities that can be expected.

Some chemical terms

Chemical name	Formula	Common name
Aluminium sulphate	$Al_2(SO_4)_3$	alum
Calcium carbonate	$CaCO_3$	chalk, limestone
Calcium hydroxide	$Ca(OH)_2$	lime (slaked, hydrated)
Calcium hydrogen carbonate	$Ca(HCO_3)_2$	calcium bicarbonate
Magnesium sulphate	$MgSO_4$	Epsom salt
Sodium carbonate	Na_2CO_3	soda
Sodium chloride	$NaCl$	common salt, brine
Sodium hydrogen carbonate	$NaHCO_3$	sodium bicarbonate

pH value

The pH value is a measure of the acidity or alkalinity of a solution, rated on a scale that is related to the concentration of the hydrogen ions present:

- pH less than 7 – indicates an acid
- pH greater than 7 – indicates an alkali.

Characteristics of natural water

Pure water has no colour, no taste or smell, and is neither acidic nor alkaline. Water can dissolve many substances and is sometimes termed a 'universal solvent'. It is rare for natural water to be chemically pure; even rainwater dissolves carbon dioxide as it falls through the air, and becomes mildly acidic. Water which flows over the surface of the land or through the ground comes into contact with many substances and takes some of these substances into solution or into suspension. The impurities and qualities that may be found in natural waters can be conveniently described under the headings described below.

Qualities of natural water

Inorganic matter
Dissolved inorganic chemicals, such as salts of calcium, magnesium, and sodium, cause *hardness* in water.

Suspended inorganic matter includes minute particles of sand and chalk, which do not dissolve in water. The particles are small enough to be evenly dispersed as a *suspension* which affects the colour and clarity of the water.

Organic matter
Dissolved organic materials usually have animal or vegetable origins.

Suspended organic compounds include minute particles of fibres, fungi, hair and scales.

Micro-organisms
Diseases are caused by small organisms such as certain bacteria, viruses and parasites. Some of these organisms can be carried by water if a supply is allowed to become contaminated. Examples include typhoid, cholera and dysentery.

Pollutants
Human activities add extra impurities to natural water, mainly in the form of waste from sewage systems and from industrial processes. Domestic sewage carries disease organisms which must not be allowed to contaminate water supplies. The detergent content of household sewage can also be high and difficult to remove.

Industrial contaminants include: cyanide, lead, and mercury, and agricultural fertilisers.

(continued)

Qualities of natural water (*continued*)

Acidity and alkalinity
Acidity in natural water is usually caused by dissolved carbon dioxide and dissolved organic substances such as peat. Acidic waters are corrosive and also cause plumbo-solvency where lead, a poison, is dissolved into the water from lead tanks or pipes.

 Alkalinity in natural water is more common than acidity. It is usually caused by the presence of hydrogen carbonates.

Hardness

Typical values of hardness
Soft water:
 0–50 mg/litre soft
 water
Slightly hard water:
 100–150 mg/litre
Hard water:
 200+ mg/litre

Some natural water contains substances which form a curd-like precipitate or scum with soap. No lather forms until enough soap has first been used in the reaction with the substances producing this 'hardness'.

 About 40 per cent of the public water supply in the United Kingdom is hard water. In general, hard water comes from underground sources or from surface water collected over ground containing soluble salts such as carbonates and sulphates; in limestone areas for example. Soft water tends to come from surface water collected over impermeable ground, such as in granite areas.

Types of hardness

Temporary hardness

Temporary hardness *is hardness that can be removed by boiling.*

Temporary hardness is usually caused by the presence in the water of calcium carbonate or magnesium carbonate. The scale or 'fur' found inside kettles is the by-product of removing temporary hardness by boiling.

Permanent hardness

Permanent hardness *is hardness in water which can not be removed by boiling.*

Permanent hardness is usually caused by the presence of salts such as calcium sulphate and magnesium sulphate.

Consequences of water hardness

Hardness in water has the advantages and the disadvantages listed below. Public water supplies are not usually treated for hardness and the suitability of an untreated supply needs to be assessed for each application of the water.

Bottles of special 'mineral water' are sold at premium prices because they contain the minerals that cause hard water.

Disadvantages of hard water

- Wastage of fuel occurs because of scale in boilers and pipes
- Deterioration and damage to boilers and pipes are caused by scale
- Wastage of soap and energy occurs before a lather forms
- Increased wear occurs in textiles which have to be washed for longer periods
- Industrial processes are affected by the chemicals in hard water
- The preparation and final taste of food and drinks can be affected by hard water.

Advantages of hard water

- Less toxic lead is dissolved from pipes by hard water
- 'Better taste' is usually a feature of hard water
- Decreased incidence of heart disease appears to be associated with hard water.

Sources of water supply

Rainfall is the original source of the water used for drinking. Part of the water evaporates from the Earth soon after it falls as rain. Part of this water drains on the surface to join streams and rivers, and part of the water percolates into the ground to feed underground supplies (see Figure 5.6).

The balance of evaporation, surface water and underground water varies with the particular climate, the district and the time of year. A typical proportion is one-third evaporation, one-third run-off, and one-third soak-in. A larger proportion of the rainwater is lost by evaporation during the summer.

Surface water
Streams, rivers
Lakes and reservoirs

Underground water
Springs and wells

Rainwater collectors
Roofs, paved surfaces

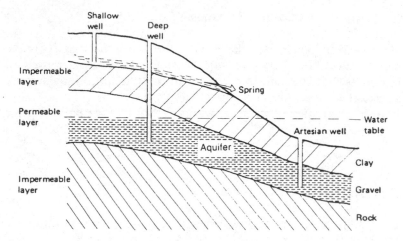

Figure 5.6   Underground water sources.

Sources of water supply are usually classified by the routes that the water has taken after rainfall. For supplies of drinking water the main categories are listed in the margin and described below.

Underground water

When rain falls on soils or porous rocks, such as limestone or sandstone, some of the water sinks into the ground. When this water reaches a lower layer of impervious material, such as clay or rock, it may be held in a depression or it may flow along the top of the impermeable layer.

Springs

A spring is a source of ground water that occurs when geological conditions cause the water to emerge naturally. The water from such a spring is usually hard with a high standard of purity achieved because of the natural purification which occurs during the percolation through the ground. All springs need protection from contamination at their point of emergence.

Wells

Wells are a source of underground water but, unlike springs, the water must be artificially tapped by boring down to the supply. Wells may be classified by the following types:

- *Shallow wells* tap water near the surface
- *Deep wells* obtain water below the level of the first impermeable layer
- *Artesian wells* deliver water under their own heads of pressure, because the plane of saturation is above the ground level.

The classification of wells as 'shallow' or 'deep' depends on the sources of water and not on the depth of their bore. Shallow wells may give good water but there is a risk of pollution from sources such as local cesspools, leaking drains and farmyards. Deep wells usually yield hard water of high purity.

Surface water

Water collected in upland areas tends to be soft and of good quality, except for possible contamination by vegetation. As a stream or river flows along its course it receives drainage from farms, roads and towns, and becomes progressively less pure. It is obviously important that the levels of pollution in rivers are controlled, and experience has shown that even rivers flowing through areas of heavy industry can be kept clean if they are managed correctly.

Water treatment

The variety of types and qualities of natural waters described in the previous section indicates that there is a wide range of substances whose concentrations may need to be adjusted before a water is used. The water for a public water supply is required to be 'wholesome', meaning that it is suitable for drinking.

The properties desirable for good drinking water are listed in the margin.

Properties for drinking water
Harmless to health
Colourless
Clear
Sparkling
Odourless
Pleasant tasting

Methods of water treatment

The principal techniques used for water treatment are described in following sections and can be summarised under the general headings given below.

- *Storage*: sedimentation and clarification.
- *Filtration*: slow sand filters, rapid sand filters and micro-strainers.
- *Disinfection*: chlorination and ozonisation.

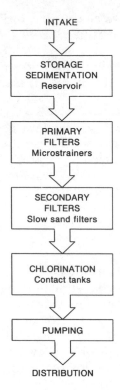

Figure 5.7   Water treatment scheme.

The methods used for the treatment of a particular water depend upon whether it is in small supplies or bulk supplies, and whether it is needed for domestic or industrial use. In Britain, the entire water supply is usually made suitable for drinking, even though most of it is used for non-drinking purposes.

Many industrial processes require water with less mineral content than is acceptable for drinking water and further treatment stages, such as softening, are then necessary. The addition of chemical compounds containing metals such as copper and aluminium needs to be carefully monitored and controlled. The components of a typical water treatment works are shown in Figure 5.7.

Water treatment components

Storage
Reservoirs are used to store reserves of water and they are also an important preliminary stage of treatment. All contaminants in the water are diluted in their effect and different

(*continued*)

Water treatment components (*continued*)

qualities of water are evened out. Pathogenic (disease-producing) bacteria tend to die when in storage because of lack of suitable food, the low temperature and the action of sunlight.

Filtration

When water is passed through a fine material, such as sand or a wire mesh, particles are removed from the water. Some filters, such as rapid sand filters, act only as a simple physical filter and the water also requires chemical treatment. Slow sand filters, however, combine a physical action with a chemical and a bacteriological action.

Slow sand filters

Water slowly percolates downwards through the sand bed which develops a film of fine particles, micro-organisms and microscopic plant life. It is this complex 'vital' layer which purifies the water by both physical and biological action. The slow sand filter is extremely effective and gives high-quality water, which needs little further treatment. These filters, however, occupy larger areas, work more slowly than other types of filter, and require cleaning of the sand.

Rapid gravity filters

The water passes downwards through the sand and the filtration is mainly by physical action. Rapid gravity filters work at a rate which is some 20 to 40 times faster than slow sand filters. Their construction is more compact than slow filters and they can be cheaper to install and to operate.

Pressure filters

Pressure filters are contained in steel pressure vessels. The construction of the sand filter inside the vessel is similar to the open rapid gravity filters and the backwashing is carried out in a similar way. The whole of the pressure cylinder is kept filled with water so that pressure is not lost during filtration and this type of filter can be inserted anywhere in a water main.

(*continued*)

Water treatment components (*continued*)

Micro-strainers
Micro-strainers are revolving drums with a very fine mesh of stainless steel wire or other material which is cleaned by water jets. Sometimes micro-strainers can produce water pure enough for sterilisation, without the need for filtration.

Disinfection
The disinfection of water supplies is intended to reduce harmful organisms, such as bacteria, to such very low levels that they are harmless. This safe quality needs to be maintained while the water is in the distribution system, including any reservoirs which store purified water.

Disinfection can be achieved by a number of agents but chlorine and ozone are usually employed to treat public water supplies.

Softening of water

Hard water is satisfactory, and even desirable, for drinking, but also has the disadvantages of causing scaling and other effects described earlier. It is not usual practice to soften public supplies of water but softening may be necessary for industrial supplies. There is a range of methods used for changing hard water into soft water, and they are summarised below.

Precipitation
Precipitation methods act by completely removing most of the hardness compounds which are present in the water. Chemicals are added to the hard water to form insoluble precipitates which can then be removed by sedimentation and filtration.

Base exchange
Base exchange methods act by changing hardness compounds into other compounds which do not cause hardness.

Demineralisation
Demineralisation is the complete removal of all chemicals dissolved in the water. This can be achieved by an ion-exchange which is a more complete process than base exchange.

Magnetic water treatment
In these treatments the hard water flows through a 'water conditioner' which applies a strong magnetic field to the water. When this water is heated, the hard water products remain as microscopic particles suspended in the water instead of forming scale. The products are then carried by the movement of the water or else form a movable sediment.

Key words for Chapter 5, Services Science

The following is a list of some keywords used in this chapter. Use the list to test your knowledge and, if necessary, consult the text to learn about the terms.

artesian well	induction	potential difference
Bernoulli	micro-organism	root mean square
electrical power	permanent hardness	value
hydroelectric	pH value	sand filter
		three-phase

6 *Materials Science*

Some materials used in construction
Timber, Papers
Concrete, Stone
Bricks, Blocks
Steel, Aluminium
Plaster, Glass
Plastics, Paints

The structures, buildings, fitting and services of the built environment use an enormous range of materials, from plain stone to microprocessor chips. The quantities required of raw materials, such as timber and stone, have a large effect on the natural environment around us.

Certain materials are chosen because of the way they perform their function, such as providing strength and durability, and their performance depends on the properties of those materials. In order to understand the properties of the materials we need to be familiar with a range of science topics which are summarised in this chapter.

The properties of materials can be examined under the following headings:

- *Physical* – such as structure, density, surface properties
- *Mechanical* – such as strength, stiffness, toughness, hardness
- *Chemical* – such as bonding between atoms, resistance to attack

Physical properties of materials

Physical properties of a material include: its structure and its visible appearance; how the material behaves with forces such as compression; and how the material behaves with energy such as heat.

Appearance

Important information about a material is gained just by looking and feeling. You don't need an advanced laboratory or qualification, you just need your senses. Consider the following observations that might be recorded.

Physical form: Lengths, sheets, blocks.
                         Lumps, powder, crystals.

Colour: Dark or light.
               Transparent or opaque.

Surface: Smooth or rough.
               Sealed or porous.
               Plain or patterned.

These properties affect the manner in which materials or buildings behave. Silver-coloured roofs, for example, absorb less radiation from the Sun and also lose less heat than dark roofs. Porous surfaces allow the penetration of water.

State of matter

All materials that we see around us exist in one of the following three states or phases of matter:

Solid: fixed volume and fixed shape.

Liquid: fixed volume, no fixed shape.

Gas: no fixed volume, no fixed shape.

At room temperature a material will exist naturally in one state, but there are fixed temperatures, called the *melting point* and the *boiling point*, where that material will change state.

For example, when heat is supplied to an ice cube it melts into water and then boils into steam. Liquid molecules can also become gas molecules by the process of evaporation, such as when a puddle of water dries away.

Density

A cubic metre of concrete is the same size as a cubic metre of foamed plastic; but there is a difference in the mass!

Density is a measure of the mass contained in a fixed volume of material.

Unit: kilograms per cubic metre (kg/m^3).

$$density = \frac{mass}{volume}$$

Density of Materials

Material	Typical density (kg/m^3)
Brick	1700
Block, lightwt concrete	650
Concrete	2400
Marble	2600
Aluminium alloy	2800
Copper	8930
Iron, cast	7150
Lead	11340
Steel, mild	7860
Timber, oak	650
Timber, spruce	600
Glass, sheet	2500
Glass, wool	50
Oil, olive	920
Water	1000
Ice	920
Air	1.29
Carbon dioxide	1.98
Water vapour	0.8

Mechanical properties of materials

The mechanical properties of construction materials affect topics such as the weight, size, strength, stiffness and hardness. This information can be used to design the correct type and size of important parts of structures, such as foundations and beams, and to predict how these parts will behave when loads are placed upon them.

The forces affecting buildings are usually the static forces which cause a change in shape. These forces are opposed by an equal and opposite reaction force within the material or the structure.

For example, if you stand on a brick wall, this load produces a downwards force which is opposed by an upwards force from the wall and the ground. If this reaction didn't exist you would move downwards! The bricks in the wall change shape under the load of your weight. Even if you can't see this deformation with your eye, it does exist and it can be measured.

Loads

A force is a vector quantity, which means it always has a direction that has to be considered. Brick, for example, is strong in one direction (compression force) but weak in the opposite direction (tension force). So brick makes good blocks to stand on but does not make good ropes to hang on.

* **Tension** is a force which tends to stretch a material. Examples: cables and ropes.
* **Compression** is a force which tends to squeeze a material. Examples: chair legs and walls.
* **Shear** is a force which tends to make the surfaces of a material slide.
  Examples: bolts and shelf supports.

Stress and strain

Strength is the ability of a material to resist a force. This strength is supplied by the internal forces which oppose the applied force. These internal forces cause *stress*. We have to specify whether the stress is tensile or compressive and to take account of the area over which the external load is distributed.

A stress in a body produces a change of shape or deformation which is measured as *strain*. This change of shape is usually a change in length, either longer or shorter depending on the type of force.

Terms and formulas for mechanical properties

Force
A *force* is an agency which causes a change in either the shape
or the motion of a body.
Unit: newton (N).

Stress
Stress in a body is measured as the force applied per unit
area. Stress is calculated by the formula:

$$\text{Stress} = \frac{\text{Force}}{\text{Area}} = \frac{\text{Load}}{\text{Area}}$$

Unit: newtons/square millimetre (N/mm^2).

The non-standard use of mm in the unit gives convenient
working figures. Stress may also be measured by the pascal
(Pa) unit.

Strain
Strain in a body is measured as the change in length
compared with the original length.

Strain is calculated by the formula:

$$\text{Strain} = \frac{\text{Change in length}}{\text{Original length}}$$

Unit: dimensionless.

The units of length cancel when divided. Strain may also be
expressed as a percentage.

Elastic modulus
Elastic modulus or Young's modulus (E) of a material
compares the applied stress with the strain produced by that
stress.

The elastic modulus is given by the formula:

$$E = \frac{\text{Stress}}{\text{Strain}}$$

Unit: newtons/square millimetre (N/mm^2).

The units for E are the same as for stress because strain, on
the bottom line, has no units.

Table 6.1   Elasticity values

Material	Elasticity modulus, E (approx. in N/mm^2)
Brickwork engineering	20 000
Concrete	28 000
Steel, mild	200 000
Timber, softwood	7 000
Timber, hardwood	12 000
Aluminium, alloy	70 000
Glass fibre	20 000

Elasticity

All materials deform when subjected to a force but not all materials return to their original shape and size. Rubber bands and steel beams, for example, do recover their original shape but putty and lead pipes do not.

Elasticity is the ability of a material to completely recover its shape and size when a deforming force is removed. This property is measured by the *elastic modulus* or *Young's modulus* (see Table 6.1).

Elasticity is also used to measure the idea of *stiffness*: stiff materials have high values for their modulus of elasticity (E). We often make a beam out of steel because steel has both strength and useful stiffness. A beam made out of plastic, for example, might have enough strength but its low stiffness would cause it to sag.

Plasticity

A plastic material is one that keeps its new shape after a deforming force is removed. Such materials are easily moulded into shape and the modern polymer materials popularly called 'plastics' show this property. A plastic plate, for example, is simply made by squeezing it in a mould.

But many other materials, including metals, have a plastic stage under high loads and for materials with notable plasticity, the following extra properties may apply.

- *A **malleable** material is one that is easily deformed to a new shape by compression such as hammering.*
  Examples: clay, lead, gold leaf.

- A **ductile** material is one that is easily deformed to a new shape by a tension force such as pulling.
  Examples: gold wire, copper wire.
- A **brittle** material is one that shatters easily.
  Examples: glass, cast iron, concrete.
- A **tough** material is one that does not shatter.
  Examples: plastic, lead.

Hardness

Some materials are better than others at resisting scratching, rubbing and other forms of wear. This property can be important for applications such as floor surfaces, tools and other surfaces exposed to moving loads.

*The **hardness** of a material is the resistance to indentation, abrasion and wear.*

The measurement of hardness involves scratching or penetrating a sample with balls or needles applied under standard conditions. The various tests include the Brinnell, Rockwell and Vickers hardness tests. The mechanical property of 'hardness' should not be confused with 'strength' or 'toughness', which have different definitions, as given above. For example, diamond is extremely hard but it can be shattered with a hammer, so it is not tough. A plastic bag is tough but it is not hard. Glass sheet is hard and strong but it is not tough.

Chemical properties of materials

The science of chemistry tells us about the ingredients contained in a material, so that we can then seek natural supplies of that material or manufacture the material out of other substances. The rules of chemistry also tell us why materials are strong or weak, how any material reacts with other materials, and how it interacts with the environment.

Elements and compounds

There are millions of different materials in the world but all materials are made up from a few basic elements. There are only

Some common elements

Element	Symbol
Aluminium	Al
Calcium	Ca
Carbon	C
Chlorine	Cl
Copper	Cu
Hydrogen	H
Iron	Fe
Lead	Pb
Magnesium	Mg
Nitrogen	N
Oxygen	O
Silicon	Si
Sodium	Na
Sulphur	S
Zinc	Zn

Table 6.2   Some common compounds

Compound	Formula	Atoms in one molecule
Sodium chloride	NaCl	1 sodium 1 chlorine
Water	H_2O	2 hydrogen 1 oxygen
Carbon dioxide	CO_2	1 carbon 2 oxygen
Calcium carbonate	$CaCO_3$	1 calcium 1 carbon 3 oxygen

92 elements which occur naturally and about a dozen more which have been produced artificially. All materials in the Universe are made from these same elements.

In the natural state, elements don't usually exist by themselves but are found combined with other elements in the form of *compounds* (see Table 6.2). Sodium chloride, for example, is a compound made only from the elements sodium and chlorine. Sodium is a reactive silver metal and chlorine is a poisonous green-coloured gas. Yet the result of their chemical combination is ordinary table salt, which obviously has different properties from its ingredients!

Chemical terms

Elements

An element is a substance which cannot be separated into anything simpler by chemical means.

Compounds

A compound is a substance containing two or more different elements which are chemically joined together to form a new material with new properties.

Atoms

An atom is the smallest part of an element which can take part in a chemical reaction.

Molecules

A molecule is the smallest part of a compound which can take part in a chemical reaction.

We sometimes need to consider the smallest 'bit' of a material. For an element, the smallest part is an *atom* of that element. We can't say an 'atom' of water or an atom of concrete because they are compounds of several elements. So we also use the idea of a *molecule* which is a group of atoms bonded together.

Mixtures

If substances are not chemically joined in a compound then they may be a mixture. The proportions of the components in a mixture can vary and the components can be separated relatively easily.

Mixtures which show these properties include sand and iron filings, and oil and water. Sand and cement powder, for example, starts as a mixture and forms chemical compounds when water is added to start the chemical reactions.

Solutions and suspensions

Some mixtures are formed by dissolving a substance in a liquid to form a solution. For instance, sugar dissolves in tea, and air dissolves in water. The *solute* is the dissolved substance and may be a solid, liquid or gas. The *solvent* is the dissolving liquid, and is often water.

$$\text{Solution} = \text{Solvent} + \text{Solute}$$

A solution has a clear, even appearance and remains unchanged with time.

A *suspension* is a liquid in which other substances are distributed but not dissolved. A suspension has a cloudy appearance and separates if it is left standing. Clay in water, for example, is a suspension and the clay settles to the bottom when left standing. An emulsion paint is a form of suspension and, eventually, the ingredients separate. This is why paint needs to be stirred thoroughly.

Chemical processes

Chemistry involves many thousands of combinations between different elements and compounds. The study and the use of chemistry are made easier by grouping substances and reactions into categories which have similar properties. The processes outlined here have a significant effect on the environment around us and the buildings we live in.

Chemical reactions

An important feature of chemistry is the making of new substances by chemical reactions.

*A **chemical reaction** is an interaction between substances in which atoms are rearranged.*

Chemical reactions are shown in shorthand by chemical equations which are written in molecules. The atoms are rearranged during a reaction but the total number of atoms, of each element, must remain the same on each side of the equation. For example

$$CaCO_3 \rightarrow CaO + CO_2$$

means that, when heated, one molecule of calcium carbonate (limestone) produces one molecule of calcium oxide (burnt lime) plus one molecule of carbon dioxide.

Acids and alkalis

Acids, bases and alkalis are important classes of chemicals with many uses. Some occur naturally, such as those in our bodies, or they may be manufactured.

*An **acid** is a substance that contains hydrogen which can be chemically replaced by other elements.*

Examples: sulphuric acid, H_2SO_4
hydrochloric acid, HCl
citric acid (in lemons).

*An **alkali** or **base** is a substance which neutralises an acid by accepting hydrogen ions from the acid.*

Examples: sodium hydroxide, NaOH
calcium hydroxide, $Ca(OH)_2$
cement powder.

Both acids and alkalis are corrosive when they are strong, but many substances are weakly acidic or alkaline when they are dissolved in water.

Minerals and salts

Salts are a large group of compounds found naturally in the environment or manufactured as useful materials and products.

*A **salt** is the product of the neutralisation between an acid and a base.*

Examples: Sodium chloride, NaCl (common 'salt')
Calcium carbonate, $CaCO_3$ (limestone or chalk)
Calcium sulphate, $CaSO_4$ (gypsum)
Copper sulphate, $CuSO_4$.

pH scale

The *pH value* indicates the acidity or alkalinity of a solution, measured on a scale from 0 to 14, that is related to the percentage of hydrogen ions present.

pH 0 – pH 7 – pH 14
Strong   Neutral   Strong
alkali   acid

Oxidation and reduction

- *Simple oxidation* is the gain of oxygen in a chemical reaction.
- *Simple reduction* is the loss of oxygen in a chemical reaction.

Oxidation on the surface of metals forms a layer of oxide. This process can be protective as in the case of aluminium, or it can be destructive as with the rust on steel.

Reduction processes are used to extract pure metals from their ores (rocks) when carbon, for example, 'attracts' the oxygen from the metal oxide. The metal oxide is reduced and the carbon is oxidised.

Electrolysis

There are a number of useful interactions between the electrical and chemical properties of materials.

Electrolysis is the production of a chemical reaction by passing an electric current through an electrolyte

For example, when an electric current is passed through water the chemical reaction causes water, to give off the gases hydrogen and oxygen from which it is made.

The electric current enters and leaves the liquid electrolyte by two electrodes called the *anode* (positive) and the *cathode* (negative).

Polymers

Modern plastics, paints and adhesives are based in a versatile group of compounds called polymers.

*A **polymer** is a substance with large chain-like molecules consisting of repeated groups*

Most polymers are made by the artificial process of *polymerisation* which combines simple compounds or monomers into larger units.

Examples: styrene → polystyrene
ethane → polyethylene (polythene)

Manufacture of materials

Sources of material

In earlier days, when transport was difficult, most buildings were made from local materials. Some towns were made of brick because there was clay nearby, other towns were made of local stone. Transport is now easier and cheaper and many products are moved around the world. Britain, for example, imports most of its structural timber.

We are also making new types of product using advanced technology. A plastic, for example, is made from oil products by controlling the chemical reactions to get the 'designer' product that we want.

The starting point for most building products is other 'raw' materials which usually occur on or under the Earth. These raw materials are usually compounds, such as a type of rock, which will need processing to produce a useful building product (see Table 6.3). These processes often have effects on our lives and our environment.

Table 6.3 Some raw materials and processes

Building product	Raw materials	Chemical content	Manufacturing processes
Structural timber	Wood from trees	Cellulose C, H, O	Sawing Drying treatment
Structural steel	Iron ore (rock)	Fe_2O_3	Blast furnace Alloying Rolling
Cement powder	Clay, chalk or limestone	Al compounds Si compounds Ca compounds	Mixing Kiln heating Grinding
Clay bricks	Clay	Al compounds Si compounds	Moulding Kiln heating
Glass	Sand	Si compounds Si, O	Mixing Melting Shaping
Plastic products	Oil	Hydrocarbons C, H, O	Chemical change Additives Shaping
Bituminous and ashphalt compounds	Natural bitumen or oil	Hydrocarbons C, H, O	Solvents Additives

Manufacture

The raw materials for a building product need various treatments before they can be used in building. Even wood taken straight from a tree does not give its best performance and needs a number of stages, including a period of drying, before it becomes good structural timber.

The elements and compounds contained in a raw material usually need to be separated and rearranged. Iron ore, for example, is a crumbly rock mainly composed of iron oxides (such as Fe_2O_3), from which we want to isolate the iron.

Most chemical reactions work best at higher temperatures so heating is a feature of many processes. The oxygen is removed from iron ore by heating the ore with coal (carbon) which likes to combine with oxygen. The manufacturing process may also include the removal or addition of other materials to improve the properties. Iron, for example, has to contain an exact amount of carbon to produce steel.

Manufacturing processes

Some of the processes used to make common building products are summarised below.

- Chemical changes at high temperature
  Examples: cement kiln
  brick kiln
  glass furnace.
- Exposure to air – oxidation, polymerisation
  Examples: plaster hardening
  paint film formation.
- Removal of oxygen – reduction
  Examples: iron blast furnace.
- Adjustment of water content
  Examples: timber seasoning.
  concrete mixing.

Finally, the material needs finishing and shaping in some way to allow use. For example, the material may need to be crushed into powder, pulled into lengths, or flattened into sheets.

Composite materials

Modern materials are often a combination of two or more simpler materials. Concrete, for example, is a composite of

Table 6.4   Examples of composite materials

Composite	Constituents	Improved properties
Modern steels	Iron Carbon Other elements	Tensile strength Ductility Hardness
Steel-reinforced concrete	Concrete Steel	Tensile strength
Glass-reinforced concrete (GRC)	Concrete Glass fibres	Improved strength and impact resistance
Glass-reinforced plastic (GRP)	Polymer resin Glass fibres	Improved strength and impact resistance
Cermet	Ceramic Metal	Hardness, strength and impact resistance

stones (aggregate) and cement paste which are locked together. The aggregate by itself is loose and cement paste by itself is weak and brittle.

*A **composite material** is a combination of materials that gives a new type of material with new properties.*

The composite material has properties which are superior to those of the individual components (see Table 6.4). Bronze, which is a composite alloy of copper and tin, made better swords than copper or tin. Glass, for example, is very strong but brittle. But if glass fibres are held in a matrix of weak but tough plastic material then we have fibreglass, which has both strength and toughness against shock.

Selection and performance of materials

The materials used to make a building and its services need careful selection. We can choose by knowing how well the material does its task, such as keeping out the rain. But it is also important to know other performance facts, such as whether the material is easy to use, how much it costs, whether it lasts well, and whether it is harmful at any stage of its life (Table 6.5).

If we choose the wrong materials for a roof or road, for example, then this mistake can cause problems to the users: it

Performance targets
Not harmful to environment
Appropriate for purpose
Appropriate appearance
Harmless to install
Harmless to use
Easy to use
Long life
Low initial cost
Easy maintenance
Low running cost

Table 6.5 Characteristics and performance of materials

Material characteristics	*Performance checklist*
Origins	
Sources	Long or difficult transport?
Extraction	Environmental consequences?
Manufacture	Energy used for manufacture?
Physical properties	
Density	Heavy or light?
Form	Lengths, blocks, or sheets?
Appearance	Clear or opaque?
Surface structure	Colour? Shiny?
Internal structure	Rough or smooth?
	Closed or porous?
	Patterns, crystals, grains?

(continued)

Table 6.5   (*continued*)

Material characteristics	Performance checklist
Chemical properties	
Composition	Element, compound or mixture?
Reactions	Chemical names?
	Other names?
Mechanical properties	
Strength	Strong or weak – tension or compression?
Stiffness	Bends easily?
Toughness	Moulds easily?
Hardness	Shatters or snaps easily?
	Hard or soft?
Degradation	
Moisture-frost attack	Absorbs moisture?
Carbon dioxide attack	Surface breaks up?
Corrosion	Rusts?
Insect attack	Rots?
Fungal attack	Changes with age?
Thermal properties	
Expansivity	Needs allowance for movement?
Insulation	Useful for energy conservation?
Electrical properties	
Conductivity	Useful for carrying current?
Insulation	Useful at high voltage?
Sound properties	
Acoustic absorption	Porous surface?
Sound insulation	Effects of mass, layers, gaps?
Fire properties	
Fuel content	Needs special storage?
Fire resistance	Resistance for how long?
Health and safety	
Safe to manufacture	Harmful fumes?
Safe to install	Harmful to skin?
Safe to live with	Special handling required?
	Safe to dispose of?
Buildability	
Availability	Local material?
Feasible to build	Local skills available?
Costs	
Initial costs	Expensive to buy? Expensive to install?
Running costs	Costly maintenance required?
Environmental costs	Raw materials and manufacture damaging to environment?

	Concrete	*Brick*	*Timber*	*Steel*
Origins	Clay, chalk or limestone	Clay	Trees	Iron ore
Chemical components	Al, Si, Ca	Al, Si	C, H, O	Fe, C
Final manufacture	On site (includes placing and curing)	Factory	Factory	Factory
Surface structure	Porous	Porous	Porous	Sealed
Physical structure	Matrix of crystals	Amorphous	Tubular	Crystalline
Compressive strength	High	High	High	High
Tensile strength	Low unless composite with steel	Low	High	High
Deformation	Stiff	Stiff	Elastic	Elastic
Mechanical shock	Brittle	Brittle	Tough	Tough
Moisture movement	Low	Low	Medium	Low
Thermal movement	Medium	Medium	Medium	High
Thermal insulation	Low Moderate if aerated	Low	Moderate	Very poor
Fire resistance	Moderate	High	Moderate	Poor
Atmospheric effects	Spalling with water-frost; carbonation with CO_2	Spalling with water-frost; rain penetration	Insect attack; fungal attack	Corrosion
Chemical effects	Sulphates; chlorides in water; alkali aggregates			Junctions with other metals – copper

Figure 6.1 Comparison of structural materials – properties and performance.

can look unpleasant, it can cost too much money, and the effects can last for a long time. Many choices involve a balance or 'trade-off' between different factors, such as effect on the environment, costs and personal preference.

The examples discussed in this chapter are commonly found but do not include every material possible for use in construction and built environment projects. However, the principles studied in this chapter do apply to all materials and you should be able to apply them to any material and assess its likely performance. See the checklist in Figure 6.1 for helpful questions.

Key words for Chapter 6, Materials Science

The following is a list of some keywords used in this chapter. Use the list to test your knowledge and, if necessary, consult the text to learn about the terms.

Alkalis	Molecules	Plasticity
Composite	Oxidation	Polymers
Elastic modules	Physical form	Stress
Hardness		

7 Site Investigation and Preparation

The *site* is the portion of land on which a building or other project will be positioned. The characteristics of the site have a large effect on the size and shape, the appearance, the technology used, the costs and the timing of the project.

It is therefore essential that the site is carefully investigated so that its features are known and measured, especially if they are less desirable features. For example, knowledge and testing of the ground beneath the site, or the sub-strata, are an important part of the site investigation as the soil must bear the loads transferred from the foundations of the proposed structure.

Some time later, when the site has been cleared of vegetation and construction is to begin, another vital aspect of site preparation is *setting out* the exact position of the building. The exact location and level for starting a structure must be fixed with great accuracy, otherwise the structure will be illegal in its position and its various parts may not fit together.

Site investigations

A general site investigation has the main objectives listed below:

- To assess the general suitability of the site for the proposed works
- To help produce a design which is adequate and economic
- To help overcome possible difficulties and delays due to ground conditions
- To predict possible changes in conditions which may affect the project
- To maximise the potential of the site.

Site investigation is a combination of processes which range from looking at published information, such as maps, to arranging laboratory tests on the soil.

Site investigation topics
Location of site
Neighbouring properties and
   conditions
History of site
Ground conditions
Access to site
Space for operations
Services available
Demolition and clearance
Local regulations
Supply of labour
Supply of materials
Local resources

Desk survey

Desk information
Topographical maps
Aerial photographs
Geological maps
Mining histories
Industrial history
Subsidence history
Refuse tipping
Hazardous waste
Flooding history
Weather information
Local information
Previous investigations

Even before the site is visited you can learn a lot by using information which is already available. New buildings and adaptations to existing buildings require planning permission from the local authorities who may have published guidelines. Some sites, for example, may be part of a protected zone for which planning permission is not currently given and many areas have trees which are protected.

The location map of a site gives distances from roads, towns, railways, water courses and other features. Large-scale maps are available for most areas including those which show the outline of existing structures. Old maps can be useful for showing the positions of old structures which have been demolished but whose remains may affect the new operations.

Geological maps and regional guides help predict the basic ground conditions of any area and the records of past mining activities are relevant for some sites. Other histories which may need to be investigated include those for quarrying activities, refuse deposits, hazardous waste deposits and old water courses.

If the site is not local for the construction firm then it is important to know what weather conditions, both average and extreme conditions, can be expected during the course of the project. The availability of local labour may also depend on patterns created by weather conditions or by local industries.

Site information
Access roads
Existing buildings
Remains of past buildings
Ponds, springs, streams, ditches
Shape of site
Slope of site
Vegetation on site
Trees on and near site
Overhead power lines
Outcrops of rock or soil
Ditches or infills
Electricity and gas supplies
Water supplies
Drains and sewers
Telephone links
Bench marks
Hedges and fences
Adjacent sites and buildings
Local residents
Prevailing weather

Site visit

A site visit is essential to check the information obtained from the desk survey and to gather new information. Plans and sketches of the site form the basis of most reports and should consider the items in the margin list.

The site visit is also an opportunity to gather facts and impressions of the area. Local residents can often provide extra information about the history of the site and the district.

Soil investigation

The *subsoil* beneath the surface of the ground has to support the loads from a building and it is important to find the position of a subsoil of suitable strength. The visible *topsoil* usually consists of decaying vegetable matter which is good for growing plants

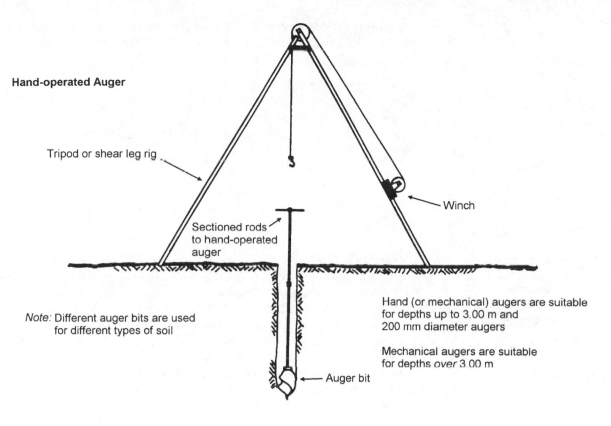

Hand-operated Auger

Tripod or shear leg rig

Winch

Sectioned rods to hand-operated auger

Note: Different auger bits are used for different types of soil

Hand (or mechanical) augers are suitable for depths up to 3.00 m and 200 mm diameter augers

Mechanical augers are suitable for depths *over* 3.00 m

Auger bit

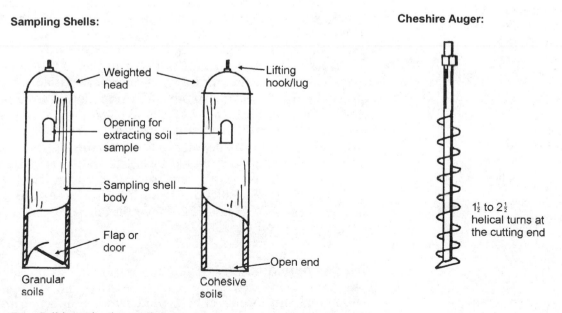

Sampling Shells:

Weighted head

Lifting hook/lug

Opening for extracting soil sample

Sampling shell body

Flap or door

Open end

Granular soils

Cohesive soils

Cheshire Auger:

$1\frac{1}{2}$ to $2\frac{1}{2}$ helical turns at the cutting end

Figure 7.1 Soil investigation equipment.

but has little strength. The subsoil can be investigated by digging trial pits and shallow trenches, or by making bore holes. A soil investigation includes the following aims:

- Define the soil type
- Find the groundwater level
- Check for tree roots
- Take soil samples for tests in a laboratory
- Carry out some *in-situ* tests
- Test for contamination.

Pits

Health and safety
Trial pits often need
  earthwork support
Trial pits often need fencing to
  prevent people falling in

For light buildings with simple foundations, such as houses, the bearing soil generally lies within two metres beneath the surface; exploration should extend below foundation level to at least 1.5 times the width of the proposed foundation. Trial pits can be dug by hand or by mechanical excavator. The pits should be made as near as possible to the proposed foundations but their backfilled soil should not interfere with the eventual works. Pits allow the subsoil to be examined undisturbed and may also help to investigate existing underground services.

Bore holes

Borings are generally used to investigate soils at depths greater than 3 metres. An *auger* is a long thin cutting tool which creates a bore hole into the ground by rotation or other force (see Figure 7.1). The cutting edge may be similar to a corkscrew or may have an arrangement to both cut and bring back a sample core when the auger is withdrawn. A *hand auger* (Figure 7.1) is a light and portable version suitable for sampling soft to firm soils near the surface. Mechanical boring uses lightweight rigs which are often fitted on the back of four-wheel drive vehicles.

Borings may be made at regular intervals or at isolated positions relevant to the proposed works. Samples of subsoil can be extracted loose (disturbed) or captured in steel tubes (undisturbed). The results are recorded on a bore-hole log as shown in Figure 7.3. Figure 7.2 shows the conventional symbols for a range of soil types.

Soil description

Soils are initially examined by eye and hand, and then described by the following properties in a standard sequence:

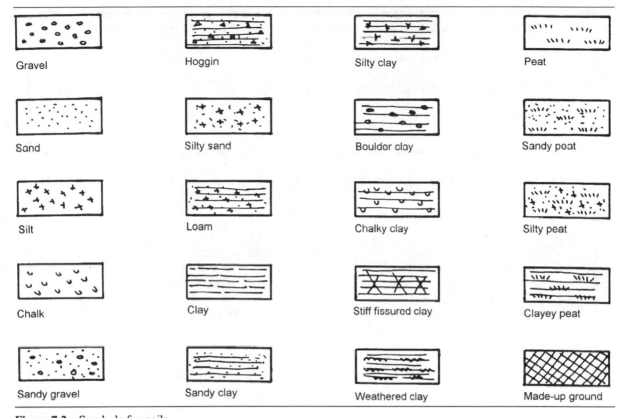

Figure 7.2 Symbols for soils.

- Consistency and structure
- Colour
- Particle size
- Soil type.

Examples of soil description

stiff fissured	dark grey	silty	CLAY
loose	brown	coarse	GRAVEL
uniform	grey	fine	SAND

Soil tests

Some soil tests can be made on site and others require taking samples that can be tested in a laboratory. In both cases, the

Site: North Street, Greentown, London				Borehole No: 1	
Type of boring: Shell and Auger				Date of bore:	
Ground level: 43·85 m above Ordnance datum				Water level: 42·35 m	
Borehole diameter: 150 mm ∅				Remarks: Borehole lined	

Description of strata	Sample	Depth	Legend	Depth [m]	Total [m]
Topsoil		0·200		0·200	0·200
Composite of clay and rubble	D	1·000		0·900	1·100
		1·500		WATER TABLE.	
Black silt with gravel	U	2·000		2·040	3·140
	U	3·000			
Coarse silt		0·050		0·050	3·190
	D	4·000			
	D	5·000			
	U	6·000			
Soft black silty clay	U	7·000		3·950	7·140

Figure 7.3 Sample bore-hole log.

Soil types
Rocks
Cohesive soils (clays, silt)
Non-cohesive soils
  (gravels, sands)
Organic (peat)
Made ground

Types of soil tests
Chemical tests
Strength tests
Compressibility tests
Shear strength
Plasticity tests

exact site position of the test sample must be carefully recorded by using a bore-hole number and position on a map of the site.

On-site testing

The following selection of tests can be carried out on site and have the advantage of leaving the soil undisturbed.

Standard penetration test assesses the density of granular soils by driving a tube into the soil with repeated drops of a standard weight.

Vane test assesses the shear strength of cohesive soils by pushing a steel blade (vane) into the soil and reading the turning force (torque) required to turn the instrument.

Loading test *assesses the likely settlement of a soil by adding known loads at regular intervals on a plate or slab and measuring the settlement.*

Standpipes *may be sunk to monitor groundwater levels over a period of time.*

Laboratory testing

A wide range of laboratory tests can be carried out on soil samples, disturbed (loose) and undisturbed, brought from the site. The tests are made using equipment in standardised repeatable routines which are detailed in documents such as the British Standards. Some common laboratory tests are summarised below.

Bulk density test *gives the weight per unit volume of a soil including the voids filled with air and water.*

Sulphate content test *for soil and groundwater. Higher levels of the sulphate group of chemicals in the soil or groundwater react destructively with concrete and mortar. Special cement powder (sulphate-resisting) therefore needs to be used in such conditions.*

Triaxial tests *for shear strength of an undisturbed sample of cohesive soil.*

Sieve tests *for the range and distribution of particle size.*

Setting out

Before operations begin on site, a point of known height or level needs to be established on the site by an engineer or technician. The height above sea level of this site point is measured from one of the national Ordnance Survey *bench marks* which are set in brick and stone at regular intervals in towns and the countryside and shown on Ordnance Survey maps. This site bench mark is then used for all levelling purposes and for checking to ensure that the building is at the correct level.

Levelling

Values for levels are established by finding the difference in height from a known level. Differences between heights are usually measured by viewing a vertical measuring *staff* through a *level* instrument, which is essentially a horizontal telescope on a tripod. The standard methods of levelling involve careful use of the equipment and making accurate records at different positions on site of rises and falls in levels compared with the bench mark.

Knowing the values of levels for a site allows calculation of the volumes of subsoil to be excavated and removed from site. The position of drains, their falls (slope) and their connections to main sewers depend on accurate levels. The vertical positions of the floors and roofs also need to be correct, especially if they are to fit against other structures.

Setting out equipment

Levelling staff: *a tall vertical pole with markings which are viewed at a distance through the telescope of a levelling instrument.*

Level or *levelling instrument*: *a surveying instrument used to establish horizontal lines, often by a telescope arrangement. Types of level include: dumpy level, automatic level, and laser level.*

Builders' square: *a timber frame in the shape of a right-angled triangle with sides between 1 and 3 metres in length. The frame can be made accurate by using simple ratios of sides, such as 3 : 4 : 5, predicted by the theorem of Pythagoras.*

Pythagoras: *a theorem in geometry which shows that some right-angled triangles have simple ratios for the lengths of their sides, such as 3 : 4 : 5. Use of these lengths must give a right-angle.*

Profile boards: *horizontal boards fixed to two upright posts in the ground just outside the line of the foundations. The upper edges are set at a known level and marked with nails or saw cuts to show the line of walls and other features.*

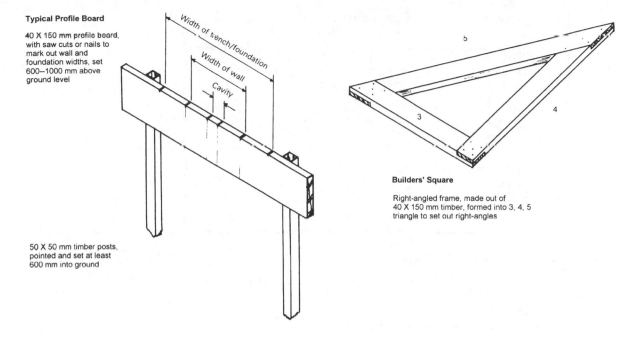

Typical Profile Board

40 X 150 mm profile board, with saw cuts or nails to mark out wall and foundation widths, set 600–1000 mm above ground level

Width of trench/foundation

Width of wall

Cavity

Builders' Square

Right-angled frame, made out of 40 X 150 mm timber, formed into 3, 4, 5 triangle to set out right-angles

50 X 50 mm timber posts, pointed and set at least 600 mm into ground

Figure 7.4 Setting-out equipment.

Setting out

The position of a building often starts by establishing a *building line*. Local authorities may specify that the building line is a minimum distance from the centre of the road or from the kerb. The *frontage line* of the building may be set on the building line or at a fixed distance from the building line and the side walls of the building are measured from the side boundaries. These positions are set in the ground using stout pegs and stretched lines.

The rear of the building is positioned in a similar way and the squareness of corners is obtained by using a builders' square and by geometrical checks such as equal diagonal measurements. When the building outline is accurately set, profile boards are fixed at the corners and wall intersections. The profile boards, which need to be clear of the string lines and leave clearance for any excavation, are then used to govern the position and width of foundations and walls. When the walls are established the profile boards are no longer needed.

Steel-framed buildings need to be set out with great accuracy as the steel stanchions and beams arrive on site in pre-cut lengths. The setting-out of roads and their curves also requires care and accuracy by a qualified team as the position of roads is

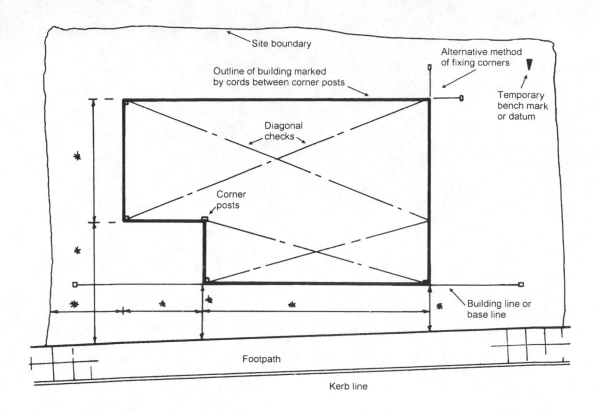

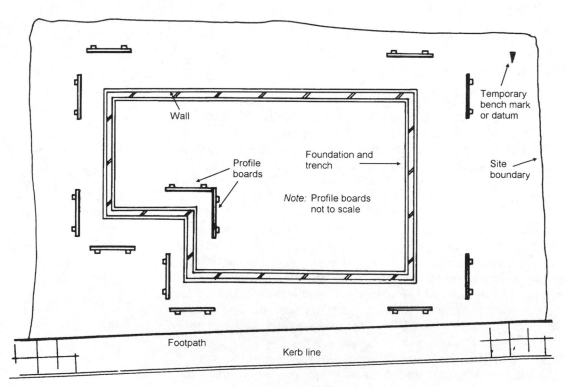

Figure 7.5 Setting-out example.

112

often used as a reference for the position of buildings and drains. The legal ownership of property depends on the accurate location of boundaries and buildings.

Key words for Chapter 7, Site Investigation and Preparation

The following is a list of some keywords used in this chapter. Use the list to test your knowledge and, if necessary, consult the text to learn about the terms.

Auger	Desk survey	Profile board
Bore hole	Frontage board	Topological map
Builder's square	Penetration test	Trial pit
Cohesive soil		

---

8 Substructures and Groundworks

The *substructure* is that part of a building or other structure which is below the ground, unlike the superstructure which is above the ground. Typical stages or elements of the substructure, such as foundations, are listed in the margin.

Foundations

The foundations of a building are that part of the structure which is in direct contact with the subsoil. The foundations have to transfer loads safely into the ground without excessive settlement, movement or damage to the building both during its construction and throughout its life.

Types of loads

Dead load: the constant weight of the structure itself.

Imposed or live loads: loads created by the weight of people, vehicles, furniture and other objects placed in or on the structure. This type of loading may vary greatly.

Other loads: maximum forces produced by snow and wind must be estimated and allowed for in the design.

The structural loads may be transferred directly to the subsoil by using simple strips on which the walls rest, or the loads may be 'spread' by using a wider pad. The pad reduces the stress on the soil because, for a constant load force, increasing the area reduces the pressure at a point. Other foundations may be designed to reach a deeper layer of subsoil which avoids effects

Force and pressure

$$\text{Pressure} = \frac{\text{force}}{\text{area}}$$

near the surface such as shrinkage of clay. In poor soils or for heavy structures, piles may be used to reach stronger soils or rock at deeper levels.

Soil behaviour

The soil on which the foundations rest must be able to resist the loads from the foundations with equal and opposite forces. The soil therefore needs to have enough *compressive strength* and *shear strength* to prevent contraction and sliding within the soil. Soil strengths can be predicted by using the results of a soil investigation as described in Chapter 7.

Cohesive subsoils, such as clay, depend on the presence of water for their internal strength and bonding, and also produce swelling and shrinkage as their water content increases or decreases. Non-cohesive soils, such as sandy soil, depend on size and compaction for their strength but can be affected by surface water eroding particles or freezing in winter.

The downwards forces from the structure are distributed in the soil in a bulb-shaped zone of pressure around the foundation.

The types of foundation used for low rise construction are summarised below. Details are also shown in Figures 8.1, 8.2 and 8.3.

Traditional narrow strip

A trench is excavated under the line of the walls and the soil at the bottom of the trench is compacted. At least 150 mm of concrete is placed at the foot of the trench, and brick or block walls are then built up to the damp-proof course which must be at least 150 mm above ground level. The trench needs to be supported while the footings are built and, when finished, the soil removed from the trench needs to be filled back in around the foundations.

Features of narrow strip foundations

- Time-consuming to construct
- Needs several skilled trades
- More materials, labour and expense than other forms.

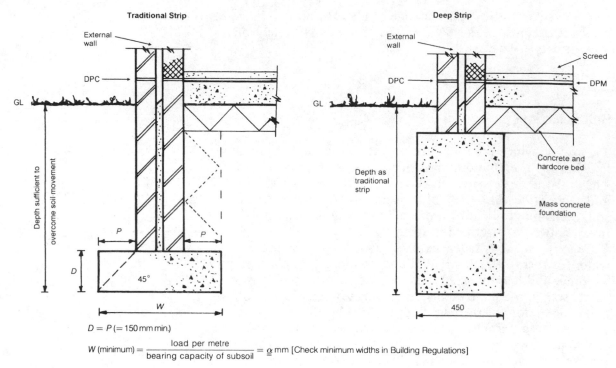

Figure 8.1 Strip foundations.

Wide strip foundations

When a soil has a low load-bearing capacity, the strip at the bottom of the foundation may be increased in width. The larger area of strip spreads the load over a larger area of soil and reduces the pressure at each point.

Deep strip foundations

A trench is excavated to the width and depth needed and then all of the trench is filled with concrete to a level which stops within two brick courses of ground level. This type of foundation may also be known as a *trench fill* foundation.

Features of wide strip foundations

- Less excavation works needed than traditional strip foundation
- Fewer skilled workers needed
- Can be cheaper than a traditional strip foundation.

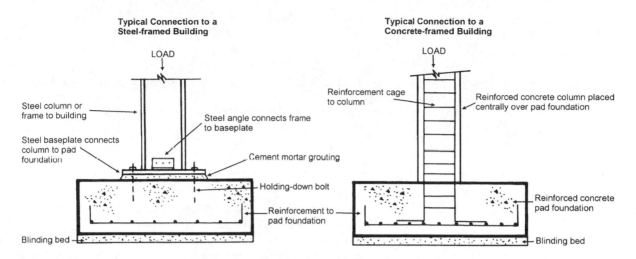

Figure 8.2   Pad foundations.

Pad foundations

Pad foundations act like a series of small isolated raft foundations. Each pad is usually square in plan shape and transfers the load from a column on to the ground. Pad foundations are commonly used to support the steel or concrete frames of retail sheds or warehouses.

Features of pad foundations

- Relatively little excavation work needed
- Position of pad must be accurate
- Relatively rapid construction time.

Raft foundations

A concrete bed is constructed which is at least equal to the base area of the building. The raft 'floats' on the subsoil, like a raft on water, and can be used for light buildings or where the top 600 mm of subsoil overlies a substrata of poorer quality. Raft foundations can be varied by thickening the edges.

Features of raft foundations

- Relatively little excavation work needed
- Minimum skilled labour needed
- Less expensive to construct than traditional strip and deep strip foundations
- Generally restricted to lightly loaded buildings.

Short-bored pile foundations

Pile holes of suitable size are bored by a powered auger and filled with mass concrete. Reinforcement may be used to resist the effects of ground movement. A reinforced concrete *ground beam* connects the individual piles heads together and brick walls are built from the top of the ground beam up to the damp-proof course level.

Features of pile foundations

- Relatively little excavation work needed
- Relatively rapid construction time
- Minimum skilled labour needed
- May be more economical to construct than traditional strip and deep strip foundations but is more expensive than a raft foundation
- Suitable for heavily loaded buildings
- Suitable for clay soils which are susceptible to movement
- Suitable for sites with trees (or where trees were recently removed).

Foundation design

Foundation selection factors
Soil conditions
Type of structure
Structural loads
Economics of foundation types
Time available for construction
Life of building
Future use of building

The decisions about which type of foundations are best for a particular structure can be considered under the following stages:

- Soil investigation
- Assessment of soil loading
- Assessment of loads
- Choice of foundation type
- Size of foundations.

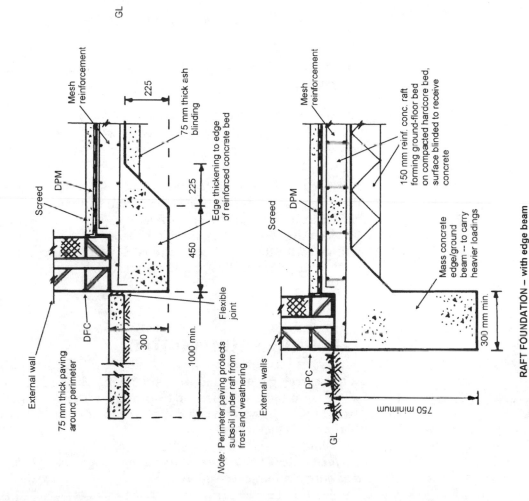

SHORT-BORED PILE

Screen

DPM

External wall

DPC

Lean mix concrete cavity fill

40 mm thick blinding concrete

Concrete and hardcore beds

Reinforced concrete ground beam on blinding concrete 40 mm th ck. Size of beam designed to suit the loadings

Actual depth of piles is governed by the loadings, the bearing capacity of the subsoil and its stability — max. economic depth 4.50 m

Bored and cast *in situ* concrete piles, 250–400 mm in diameter

Actual size and spacing of piles are designed to suit loadings

Typical spacing — 1.80 to 2.50 m c/c

Typical loading — 40 to 125 kN per pile

Bored piles formed by lorry- or tractor-mountec auger, up to 80 piles per day can be bored

GL

RAFT FOUNDATION – with edge thickening

Mesh reinforcement

Screen

DPM

External wall

DPC

75 mm thick paving around perimeter

225

75 mm thick ash blinding

225

450

Edge thickening to edge of reinforced concrete bed

Flexible joint

300

1000 min.

Note: Perimeter paving protects subsoil under raft from frost and weathering

RAFT FOUNDATION – with edge beam

Mesh reinforcement

Screen

DPM

External walls

DPC

GL

750 minimum

150 mm reinf. conc. raft forming ground-floor bed on compacted hardcore bed, surface blinded to receive concrete

Mass concrete edge/ground beam – to carry heavier loadings

300 mm min.

Figure 8.3 Raft foundations and short-bored piles.

119

Earthwork support

Holes and trenches need to be dug in the ground for foundations and drains. There are natural forces in the earth which try to close these gaps in the ground so that excavations and earthworks are dangerous areas for site operatives unless they are supported correctly.

There are various systems for supporting earthworks and most of them use timber planks or sheeting to resist the forces of the earth. In the case of a trench, most of the forces exerted by one side of the trench is balanced against the forces exerted by the other side.

Timber is a versatile material for supporting earthworks in shallow excavations but steel is also used for the sides of supports and the struts which run between the side of a trench.

Poling boards *are vertical boards used to support the sides of an excavation.*

Waling boards *are horizontal boards used to support the sides of an excavation.*

Struts *are horizontal members in earthwork supports which connect poling or waling boards on opposite sides of a trench.*

The choice of system for supporting excavations depends on the type of soil which is being excavated. Compact soils, such as clay, can be retained by *open timbering* where there are spaces between the shoring. Dry mobile soils, such as sand, need to be retained with the use of continuous sheeting made of timber or steel. Examples are illustrated in Figure 8.4.

Ground-water control

When carrying out ground works there may be a need to remove water from the excavation or to prevent water getting near the excavation. The water being removed can be classified as surface water or subsoil water.

Surface water *is rain water which is on the surface of the ground and may run into excavations.*

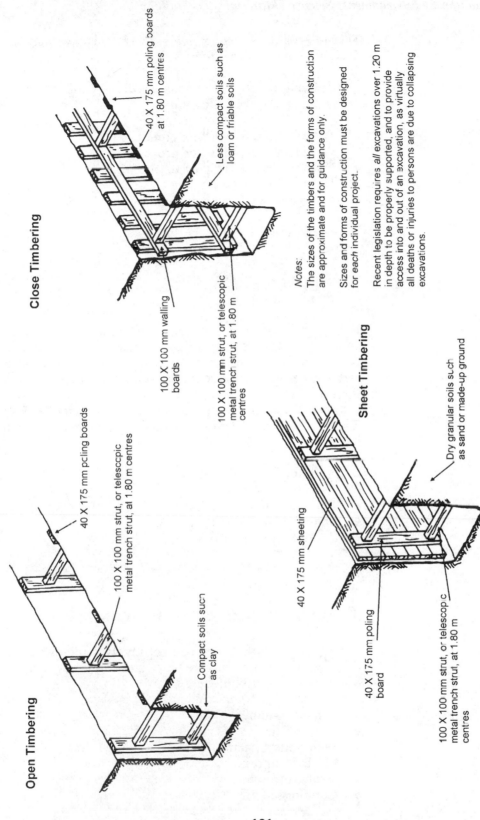

Open Timbering

40 X 175 mm pcling boards

100 X 100 mm strut, or telescopic metal trench strut, at 1.80 m centres

Compact soils such as clay

Close Timbering

40 X 175 mm poling boards at 1.80 m centres

Less compact soils such as loam or friable soils

100 X 100 mm walling boards

100 X 100 mm strut, or telescopic metal trench strut, at 1.80 m centres

Notes:
The sizes of the timbers and the forms of construction are approximate and for guidance only.

Sizes and forms of construction must be designed for each individual project.

Recent legislation requires *all* excavations over 1.20 m in depth to be properly supported, and to provide access into and out of an excavation, as virtually all deaths or injuries to persons are due to collapsing excavations.

Sheet Timbering

Dry granular soils such as sand or made-up ground

40 X 175 mm sheeting

40 X 175 mm poling board

100 X 100 mm strut, or telescopic metal trench strut, at 1.80 m centres

Figure 8.4 Earthwork support.

121

Subsoil water is the water which is to be found within the subsoil. It is free-moving when it is above the water table level.

Removing water is known as *dewatering* and is carried out in order to achieve the following:

- To provide safe working conditions
- To reduce ground-water pressure
- To increase the angle of repose of the subsoil
- To reduce humidity in the excavations
- To comply with current building regulations.

Angle of repose
The natural angle at which a material will rest

Ground-water control can be classified into the following two areas:

1. Temporary
2. Permanent.

Both temporary and permanent ground-water control can be considered under two types of remedy. The ground water may be lowered below the level of the excavation base by means of a water-lowering technique, or prevented from travelling towards the excavation by a water-exclusion barrier.

Some water-control techniques are illustrated in Figure 8.5 and factors to consider in the selection of a technique include the following:

- Amount of running surface water
- Amount of subsoil water
- Existence of a water table
- Shape of the excavation
- Subsoil strata and type.

Temporary ground-water control

Temporary ground-water control involves removing water for the duration of substructure work and can take the form of any of the following techniques:

- Sump pumping
- Sheet piling
- Well point system
- Electro-osmosis
- Ground freezing
- Compressed air.

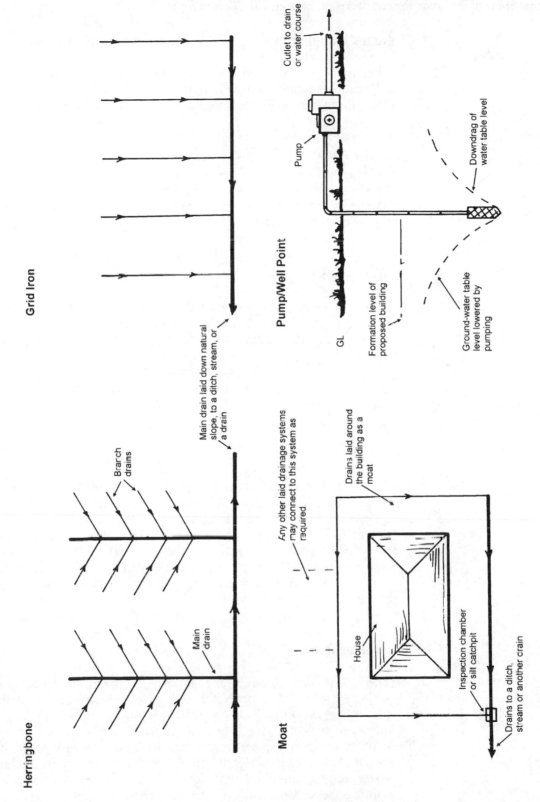

Grid Iron

Pump/Well Point

Cutlet to drain or water course

Pump

GL

Formation level of proposed building

Downdrag of water table level

Ground-water table level lowered by pumping

Main drain laid down natural slope, to a ditch, stream, or a drain

Herringbone

Branch drains

Main drain

Any other laid drainage systems may connect to this system as required

Drains laid around the building as a moat

House

Inspection chamber or silt catchpit

Drains to a ditch, stream or another drain

Moat

Figure 8.5 Ground-water control.

123

Sump pumping

This involves providing a small sump close to the general excavation, which is lower than the base of the excavation. A pump is placed in the sump and the water pumped away.

Sheet piling

This is where steel sheets are driven into the ground to provide a barrier to the ingress of water. The sheet piles may also support the side of the excavation.

Well point system

This is a system for lowering the water table level to a point below the excavation base. A number of variants of the system can be used, such as shallow bored wells, multi-stage well points and deep-bored wells. A number of wells are sunk into the ground and the individual wells connected to a header pipe which forms a ring main around the excavation from which the extracted water is pumped to an outfall.

Electro-osmosis

This is a method which can be used to dewater cohesive soils. A positive electrical charge is applied to the pump inlet filter and a negative electrical charge to the soil using a metal rod. Both the inlet filter and metal rod are inserted into the ground. Water will flow from the negatively charged rod to the positively charged inlet and can be pumped away. The system speeds up the process of dewatering but is expensive because of the large amount of electricity required.

Ground freezing

The water suspended in the subsoil is frozen by circulating a chilled solution of brine through pipes which are connected to a refrigeration unit. The principle is similar to that used in a domestic fridge. Extraction of heat from the surrounding soil reduces the soil temperature and causes the subsoil water to freeze. This frozen water forms a barrier and prevents further water movement. The technique also has the advantage of making the subsoil stable for excavation.

Compressed air

The air in the excavation is pressurised to prevent water getting into the excavation. This technique is frequently used in tunnelling work and involves the use of air lock doors and decompression chambers. The higher air pressure inside the tunnel prevents the water from seeping into the tunnel. It is more common today for the cutting heading on a tunnelling boring machine to be pressurised, rather than the entire tunnel.

Permanent ground-water control

Permanent water control involves permanently removing water from the soil and can thus reduce the risks of water penetration and dampness in the completed structure.

The techniques used include the following:

- Land drains
- Sheet piling
- Slurry cut-off walls
- Grouted membranes
- Secant piled walls.

Land drains

A series of land drains can be placed in an area to be drained in a number of configurations (see Figure 8.5):

- Herringbone
- Grid iron
- Moat
- Fan.

The drains are laid on a fine granular bed and surrounded by a coarse granular filter. The drainage pipes may be earthenware pipes of 300 mm length and 100 mm diameter, or perforated PVC pipes in lengths of 3000 mm. The lateral drains are connected to a main drain which discharges to an outfall such as a stream or river.

Sheet piling

This technique is similar to that used for temporary water control.

Slurry cut-off walls

This involves placing a cement or concrete wall slurry into the ground which will form a barrier to the movement of water. The cement slurry or concrete slurry wall is placed *in situ*.

Grout membrane

A thin grout membrane is injected into a void in the ground which has been formed by the extraction of a universal column steel section which was previously driven into the ground. The principle of preventing ground-water movement is similar to that of slurry cut-off walls.

Secant piled walls

Circular concrete piles are placed *in situ* in the ground around the perimeter of the excavation. Each alternate pile is placed and, when the first piles have achieved sufficient strength, the second series of piles are bored. The process allows the piles to become interlinked, thereby forming a continuous wall of piles around the perimeter of the proposed excavation. The method has the advantage of using the piles as foundations for the subsequent superstructure while also providing stability to the retained material outside the excavation.

Ground-water control

System	Soil type	Comment
Temporary ground water		
Sump	Clay, medium gravel, boulders, fissured rock	Only used for excavations up to 5 metres deep
Sheet piling	Most soil types except boulder beds	Can be either temporary or permanent. Will need sump pumping from excavation as piles not totally water-proof

(continued)

Ground-water control (*continued*)

System	Soil type	Comment
Temporary ground water		
Well point	Non-cohesive soils, fine silt, medium silt, coarse silt, fine sand, medium sand, coarse sand, fine gravel, medium gravel	Depth of 5 to 6 metres, deeper excavations require multi-stage well points if space available
Electro-osmosis	Cohesive soils: clay, fine silt medium silt, coarse silt	Expensive and not very common
Freezing	Medium silt, coarse silt, fine sand, medium sand, coarse sand	Suitable for excavating deep shafts and tunnels
Permanent ground water		
Land drains	All types	Generally cheapest method for large areas over shallow depths
Sheet piles	Same as temporary water control	Same as temporary water control
Slurry cut-off walls	Silts, sand and gravels	Walls do not resist earth pressure
Thin grouted membranes	Poor ground, clays silt and sand	Similar to slurry cut-off walls
Piling	All types	Piles form foundations for superstructure, deep basements in confined areas

Key words for Chapter 8, Substructures and Groundworks

The following is a list of some keywords used in this chapter. Use the list to test your knowledge and, if necessary, consult the text to learn about the terms.

Cohesive	Narrow strip	Secant piling
Electro-osmosis	Pad	Subsoil water
Grout membrane	Piled	Sump pumping
Herringbone	Raft	Waling boards

9 Structures and Superstructures

The early structures of the world, such as pyramids, buildings, bridges and towers, were designed by using experience, intuition and rules of thumb. Many of the structures collapsed or needed support, so we now use theory and mathematics to help design and build structures which are efficient and safe.

This chapter gives a general introduction to structural forms and then concentrates on building superstructure.

Structural forms

The general principles of structural theory do not depend on mathematics and the examples in this section illustrate the general principles of structures. Chapter 6 gives further details of the principles of materials and their performance.

Structural members

Structural members usually take the form of columns, beams, slabs, frames and arches. The function of a structural member is to:

- Resist loads imposed upon it
- Transfer loads to the foundation
- Provide vertical and lateral stability to the structure
- Resist bending.

*A **beam** is a* horizontal *structural member resting on two or more supports.*

Beams can be of the following types:

Simply supported – resting on two supports;
Fixed – with ends firmly built in;
Cantilever – firmly built in at one end and free at the other end;
Continuous – resting on three or more supports.

*A **Column** is a* vertical *structural member designed to transmit imposed loads to other members or to the foundations.*

Frames

*A **frame** is a composite structural member made up of a number of individual members called struts and ties.*

Individual members of a frame are named struts or ties, depending on the type of force which acts upon them:

- *Struts* are subject to *compression*
- *Ties* are subject to *tension.*

The structural member will have to behave in such a manner which avoids instability due to the forces acting on it. The forces or *loads* acting on the member may place the member in compression, tension, or torsion (twisting). The ability of both columns and beams to resist loads depends on factors which include the distance between supports, the cross sectional area, and the shape of the structural member.

Compression *is a force which tends to squeeze a material.*

Tension *is a force which tends to stretch a material.*

Shear *is a force which tends to make surfaces of a material slide.*

Types of loads
Dead load
Live/imposed load
Wind load

Beam design

A beam which is simply supported at each end and which has a point load in the centre will tend to bend or deflect, and the direction of this bending is away from the direction of the applied load. Therefore the parts of the beam furthest away from the applied load will stretch and be in tension, while the parts of the beam nearest to the applied load will push together and be in compression.

The maximum bending will take place at the centre of the simply supported beam. If the applied load 'bends' the beam so that it exceeds its maximum resistance to bending moments then it will fail and collapse.

An increase in the depth of the beam will increase its resistance to bending. The depth of the beam together with the distance

Load

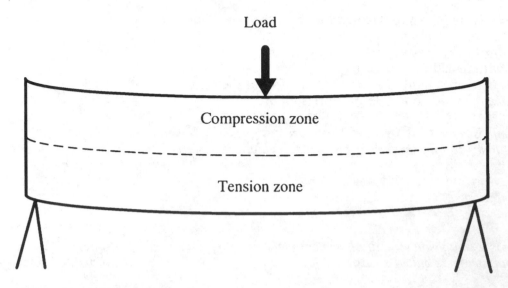

Compression zone

Tension zone

Figure 9.1   Effect of load on a beam.

between the beam supports will have a direct bearing on its ability to carry loads. This is known as the *span/depth ratio*.

Column design

A column which is subjected to a load must resist bending (buckling) in a number of directions, and this resistance can be achieved by stiffening the column. Greater stiffness can be obtained by increasing the column's effective plan width in both directions. For example, a thin tall column will buckle easily with a small load applied to it, but a 'fat' short column will resist buckling even when a greater load is applied to it. In other words, a slender column is easier to buckle.

An important structural concept in relation to columns is the *slenderness ratio*, that is, how thin the column is in relation to its height.

Load transmission

In order to design structures safely, the load path of the structure must be understood. Each load applied to the structure must ultimately be transferred safely to the foundations of the structure and thus to the ground.

The transfer of the load from one structural member to another is crucial and therefore the connection between members is important. The connections between say steel members can be by any of the following methods:

- Bolting
- Riveting
- Welding
- Cleats
- Reinforcement.

In concrete structures which are cast *in situ* (made on site), the safe transference of the loads is achieved by continuing the steel reinforcement from one member to another and tying each member together.

Load-bearing walls transfer the loads by having individual bricks or blocks bonded together by mortar or by the arrangement of the bricks or blocks.

Bond types
English bond
Flemish bond
Stretcher bond
English garden wall
Dutch bond
Monk bond

Structural forms

Structural forms can be classified into a number of basic forms such as:

- Mass
- Frame
- Shell.

Mass structures

The mass form consists of mass wall construction: a solid wall which transmits the loads on the building to the foundations along the entire length of the wall.

The mass structural form can be divided into the following two types:

- Cellular
- Cross-wall.

The cellular form encloses the entire space with the outer walls being load bearing and the inner walls a combination of load bearing and non-load bearing. A typical example of a cellular mass wall structure is a single- or two-storey detached house.

The crosswall structural form consists of a series of parallel walls and is most commonly seen in terraced houses. Crosswalls can be built up to four/five storeys high. The front and back of the terrace house is then enclosed with cladding such as timber, metal or tiles, which are hung on a non-load-bearing frame between the crosswalls.

To prevent the crosswall collapsing like dominoes, rigidity is obtained by either tying the floor to the crosswall, by increasing the thickness of the crosswall, by constructing the ends of the crosswall in a 'T' shape, or by a combination of all three.

Mass structural forms are generally constructed from bricks, blocks, or concrete. The number and positioning of openings in the walls are controlled by the Building Regulations so as not to compromise the stability of the wall.

Frame structures

A framed structure consists of interconnecting beams and columns. The frame forms the 'skeleton' of the building which is subsequently filled in with either cladding hung on the frame or with infill panels placed between the external beams and columns.

A framed structure has the advantage over mass structures in that it is lighter and capable of achieving greater heights.

Materials which can be used for the framed structure include: steel, reinforced concrete, timber and aluminium. The choice of material for a framed structure will depend on a number of factors such as:

- Plan shape of the building
- Height of the building
- Availability of specialist labour
- Cost
- Need for fireproofing
- Availability of material
- Speed of erection
- Maintenance
- Repetition of beam and column shape and size.

Framed structures can be either single-storey or multi-storey. Some single-storey framed buildings are constructed from portal frames, which have the advantage of large spans giving a clear interior uninterrupted by columns. Typical examples of portal framed buildings are single-storey warehouses and swimming pool buildings.

The beams of framed structures transmit the loads of the building to the columns which in turn transmit the loads to the pad or piled foundations. See also preceding Chapter 8 on Substructures and Groundworks.

Materials for frames
Steel
In situ reinforced concrete
Precast reinforced concrete
Timber
Aluminium

Beam and column connections
Bolts
Welds
Reinforcement
Cleats

Typical examples of framed structures
Timber-framed housing
Portal-framed sheds
Multi-storey office blocks
Multi-storey flats
Multi-storey car parks
Glass houses

Shell structures

Shell structural forms are generally used for roof types such as domes and barrel vaults, but can also be used to form the entire building. Typical materials used in shell structures include:

- Reinforced concrete
- Steel
- Plastic
- Rubber
- Canvas
- Timber.

The shell structure can be identified by the fact that the entire shell is primarily a structural element. The strength of any particular shell is affected by its geometrical shape and form.

Typical shell structures
Dome – rotation, pendentive, translational
Barrel vault – single, double, northlight
Hyperbolic paraboloids – saddle roof

Superstructure

The *superstructure* is that part of the building which is mainly above the ground, unlike the substructure which is mainly below ground. The principal *elements* or parts of the superstructure, such as walls, are listed in the margin. You can easily see the effects of the superstructure, although some parts may be covered by a *cladding* or *finish* such as roof tiles, wall plaster, or floor tiles.

All parts of a building have a purpose. The superstructure forms the general protective barrier between ourselves inside and the climate outside.

The superstructure must also remain stable and be strong enough to carry loads (forces) and transfer them elsewhere. For example, the main walls must support their own weight (dead loads) and often they must also carry the loads from the roof and the floors. These forces are transferred to the foundations and then to the ground. Other loads on the superstructure are generated by the items we place inside the building (superimposed loads) and by sideways (lateral) effects such as the wind.

Structural elements
Walls
Floors
Doors and windows
Roofs

Superstructure functions
Exclude rain and snow
Exclude damp
Resist wind forces
Reduce heat loss
Reduce heat gain
Provide security
Admit daylight
Give views out
Allow natural ventilation
Reduce noise transfer

External walls

The external walls of a building have a number of purposes. For low rise buildings, such as a house, the walls are used to support

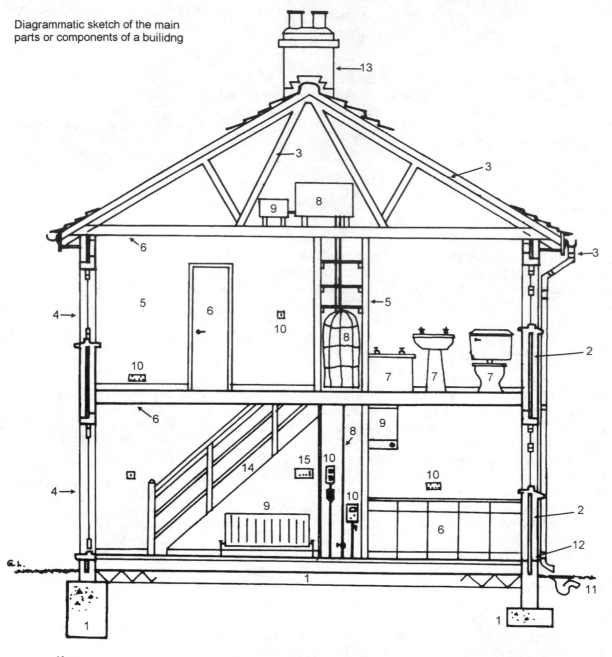

Diagrammatic sketch of the main parts or components of a builidng

Key:

1. Foundations (deep strip and traditional shown)
2. External walls
3. Roof structure, coverings and rainwater goods
4. External doors and windows
5. Internal walls/partitions and finishes
6. Floors, ceilings, staircases and joinery
7. Sanitary fittings
8. Hot and cold water supplies/distribution
9. Heating and ventilating services
10. Electricity and gas supplies/distribution
11. Drainage
12. Damp-proof course (DPC)
13. Chimney stack
14. Staircase
15. Thermostat

Figure 9.2 Parts of a building.

134

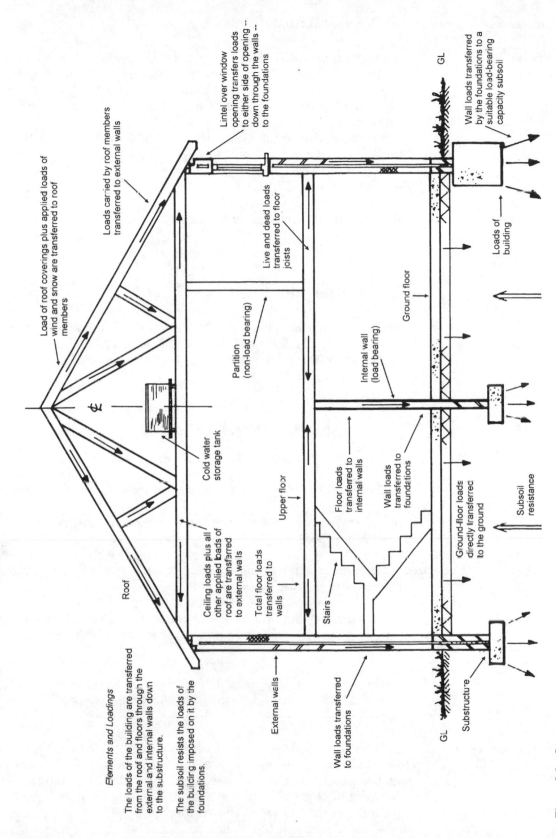

Figure 9.3 Superstructure.

Elements and Loadings

The loads of the building are transferred from the roof and floors through the external and internal walls down to the substructure.

The subsoil resists the loads of the building imposed on it by the foundations.

Load of roof coverings plus applied loads of wind and snow are transferred to roof members

Loads carried by roof members transferred to external walls

Lintel over window opening transfers loads to either side of opening -- down through the walls -- to the foundations

Wall loads transferred by the foundations to a suitable load-bearing capacity subsoil

GL

Loads of building

Live and dead loads transferred to floor joists

Ground floor

Partition (non-load bearing)

Internal wall (load bearing)

Roof

Cold water storage tank

Upper floor

Floor loads transferred to internal walls

Wall loads transferred to foundations

Subsoil resistance

Ceiling loads plus all other applied loads of roof are transferred to external walls

Total floor loads transferred to walls

Stairs

Ground-floor loads directly transferred to the ground

External walls

Wall loads transferred to foundations

GL

Substructure

135

Wall functions
Support for roof
Support for floors
Support for cladding
Rain resistance
Damp resistance
Thermal insulation
Sound insulation
Fire resistance
Resist wind load
Hold windows and doors
Good appearance

the loads from the roof and the floors. Shelter from the weather is also fundamental and can be planned by knowing the technical properties of water resistance and thermal insulation. Doors and windows need to fit into the walls and give an effect which is pleasing to look at.

The basic choices of wall construction for low rise structures are:

- Brick, block or stone
- Timber frame with cladding
- Steel or concrete frame with panels.

Brick and block walls

Bricks and blocks have to be laid one on top of each other to form a wall. The following terms and definitions are based on bricks but many of the terms also apply to the use of concrete blocks and stone blocks.

Brickwork and blockwork terms

Courses: bricks and blocks are laid in rows called courses.

Mortar is used to hold the individual bricks and blocks together.

A stretcher is a brick laid so that its long side runs parallel to the face of the wall.

A header is a brick laid so that its end or short side is laid parallel to the face of the wall.

Wall-ties are lengths of metal or plastic used to connect across a cavity and join the inner and outer leaves of brick or block wall.

Bonding is the technique of laying bricks on top of one another in a pattern so that one brick always overlaps another.

Quoin: an outer corner of a building

To achieve good *bonding*, the bricks are arranged so that joins between bricks or blocks are not continuous. This overlap distributes loads over a large area and helps stop movement between bricks. Different methods of bonding result in different patterns in the bricks, as shown in Figure 9.4.

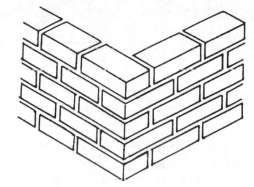

Stretcher Bond -- quoin detail

English Bond -- quoin detail

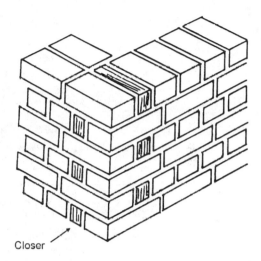
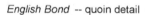

Closer

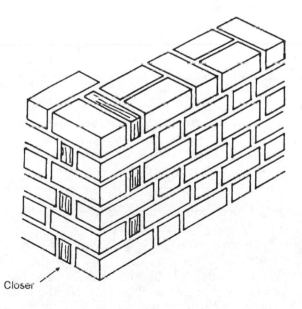

Flemish Bond -- quoin detail

Note: Closer bricks are cut bricks used to maintain the bond at corners and junctions of walls

Closer

Figure 9.4 Brick bonding.

Types of bonding

English bond
English bond is arranged as follows:

First course:	stretcher stretcher stretcher stretcher
Second course:	header header header header
Third course:	stretcher stretcher stretcher stretcher
	Pattern then repeats

English bond is a strong bond often found in one-brick thick walls, such as party walls between houses.

Flemish bond
Flemish bond is arranged as follows:

First course:	header stretcher header stretcher
Second course:	stretcher header stretcher header
Third course:	header stretcher header stretcher
	Pattern then repeats

Flemish bond gives an attractive pattern and is often seen in the walls of houses built before the 1930s. These walls are usually one-brick thick and have no cavity.

Stretcher bond
Stretcher bond is arranged as follows:

First course:	stretcher stretcher stretcher
Second course:	stretcher stretcher stretcher
Third course:	stretcher stretcher stretcher
	Pattern then repeats

Stretcher bond gives a leaf of half-brick thickness which is seen in the outer face of most modern houses. Stretcher bond is used for both leafs of an all-brick cavity wall and may be used for the outside leaf when there is blockwork or a timber frame behind the cavity. A half-brick wall does not have good stability by itself and needs wall-ties to link the outer brick leaf to the inner structure of the wall.

Typical external wall construction

The headings below give details of common wall constructions in houses in the British Isles. The 'layers' found in a wall are listed in order from outside to inside. See also Figure 9.5 for details of typical sizes and assemblies.

Traditional solid brick wall

- Brick leaf – Flemish or English bond
- Plaster.

Modern brick cavity wall

- Brick outer leaf – stretcher bond
- Air cavity – with wall-ties
- Insulation – mineral wool or expanded plastic
- Concrete block inner leaf
- Plaster or plasterboard dry lining.

Timber frame wall

- Outer brick leaf
- Air cavity – with wall-ties and fire stops
- Plywood sheathing
- Timber frame – with insulation
- Vapour barrier
- Plasterboard.

Internal walls

Internal walls or partitions are used to divide the floor area of a building into separate areas or rooms. The walls usually provide privacy from view and some sound insulation. Sometimes the walls may need to transmit loads from floors or roof above to a suitable foundation below. Some internal walls may also need to have a specified fire resistance.

Internal wall constructions
Brick or concrete block
Frame with plasterboard
Frame with timber panels
Demountable panels

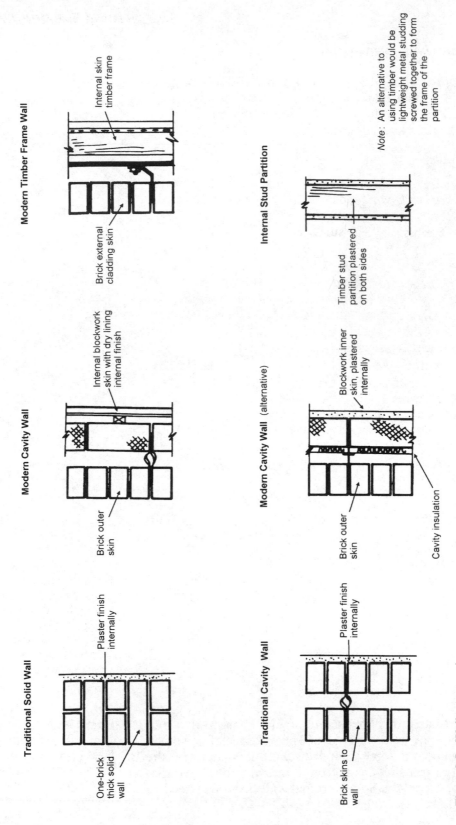

Modern Timber Frame Wall

Internal skin timber frame

Brick external cladding skin

Internal Stud Partition

Timber stud partition plastered on both sides

Note: An alternative to using timber would be lightweight metal studding screwed together to form the frame of the partition

Modern Cavity Wall

Internal blockwork skin with dry lining internal finish

Brick outer skin

Modern Cavity Wall (alternative)

Blockwork inner skin, plastered internally

Brick outer skin

Cavity insulation

Traditional Solid Wall

Plaster finish internally

One-brick thick solid wall

Traditional Cavity Wall

Plaster finish internally

Brick skins to wall

Figure 9.5 Wall construction.

140

Typical internal wall construction

The headings below give details of common types of internal walls and partitions. See Figure 9.5 for details of typical sizes and assemblies.

Concrete blocks
Lightweight concrete blocks are used to construct internal walls which do not have to carry loads. The blockwork needs to be tied to the adjacent walls and structures. Plaster can be used to fill in irregularities and give the surface for the final finish.

Timber frame with plasterboard
A timber frame of *studs* and horizontal *noggins* is built between top and bottom plate beams attached to the ceiling and the floor. Standard spacings in the frame are used to make it easy to attach the plasterboard or other cladding.

Demountable partitions
A wide range of systems are available for dividing commercial spaces such as offices. A variety of materials are used to make the partitions and they may include glazed areas. If the space needs to be rearranged then the partition can be taken down (demounted) and used again.

Floors

Floor constructions provide a level base surface for a room and also make a ceiling for any room below the floor. A basement floor provides a barrier between a room and the ground so it must resist damp, water and include thermal insulation. Sound resistance may also be important, especially in floors between new flats or in houses converted into flats.

The basic choices for modern floor construction are listed below.

Floor functions
Level surface
Strength and stability
Damp resistance
Thermal insulation
Sound insulation
Fire resistance

Timber floors

- Suspended timber ground floor
- Timber upper floor
- Floating floor.

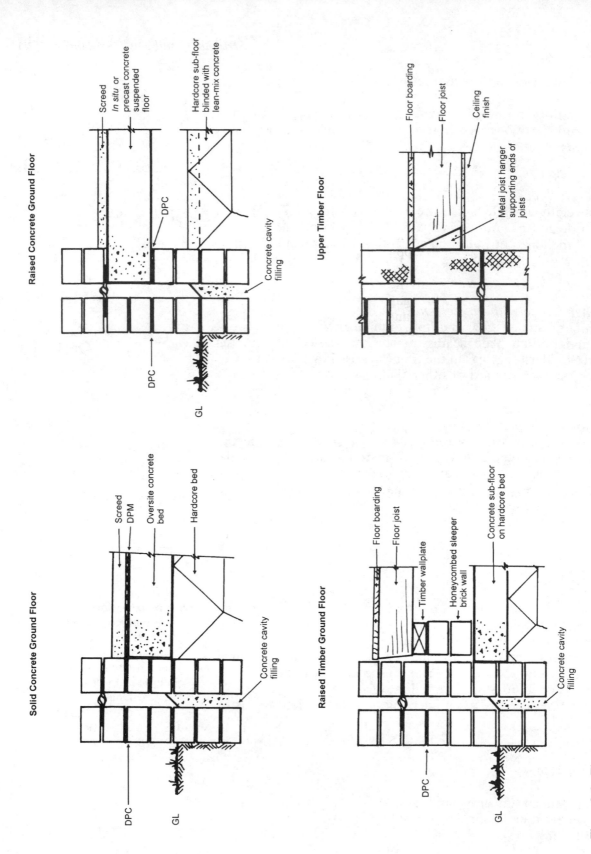

Raised Concrete Ground Floor

Screed

In situ or precast concrete suspended floor

Hardcore sub-floor blinded with lean-mix concrete

DPC

Concrete cavity filling

DPC

GL

Upper Timber Floor

Floor boarding

Floor joist

Ceiling finish

Metal joist hanger supporting ends of joists

Solid Concrete Ground Floor

Screed

DPM

Oversite concrete bed

Hardcore bed

Concrete cavity filling

DPC

GL

Raised Timber Ground Floor

Floor boarding

Floor joist

Timber wallplate

Honeycombed sleeper brick wall

Concrete sub-floor on hardcore bed

Concrete cavity filling

DPC

GL

Figure 9.6 Floor types.

142

Concrete floors

- *In-situ* ground floor
- *In-situ* upper floor
- Precast floor systems.

A *solid* ground floor rests on the ground at all points, or rests on layers of compacted material which is in contact with the ground. A solid concrete floor on the ground must contain a damp-proof membrane which may be under the solid concrete layer, or between the concrete layer and the surface screed. A *suspended* ground floor is supported by selected points and does not touch the ground between these points.

Upper floors and flat roofs have no intermediate resting points. The floor must span the gap between load-bearing walls while supporting its own weight and any loads placed on it. The structure of the floors must be designed to resist sheer stresses which are at a maximum close to the wall and to resist bending effects which are at a maximum in the centre of the span.

The size and thickness of timber joists and beams or reinforced concrete beams and slabs depend on the span between the supports, the total load expected, and the spacing of the timber or steel reinforcement. Various formulas or tables are used to find a suitable size. In general the depth of the floor joist or beam must increase to span a greater distance.

Roofs

The roof of a building provides the main barrier against the weather. The list of roof functions shows how the roof has to perform many functions during the course of its life, when it might be exposed to cold, frost, snow, hot sunshine and pouring rain.

The many different designs of roof have produced a variety of roof shapes, styles and constructions with special terms to describe them. See the list of roof terms and the labels on Figures 9.8 and 9.9 for details.

The basic types of roof construction can be classified by their shape:

- Flat roof
- Lean-to roof
- Pitched roof.

These basic roof units may be combined to produce multiple roofs. The combinations produce *gable ends*, *valleys*, *hips* and other shapes which are shown in Figure 9.8.

Roof functions
Rain resistance
Wind resistance
Load resistance
Thermal insulation
Resistance to solar heat gain
Resistance to condensation
Fire resistance
Accommodation of
  movement
Durable finish
Adequate drainage of rain
Easy maintenance
Long life
Good appearance

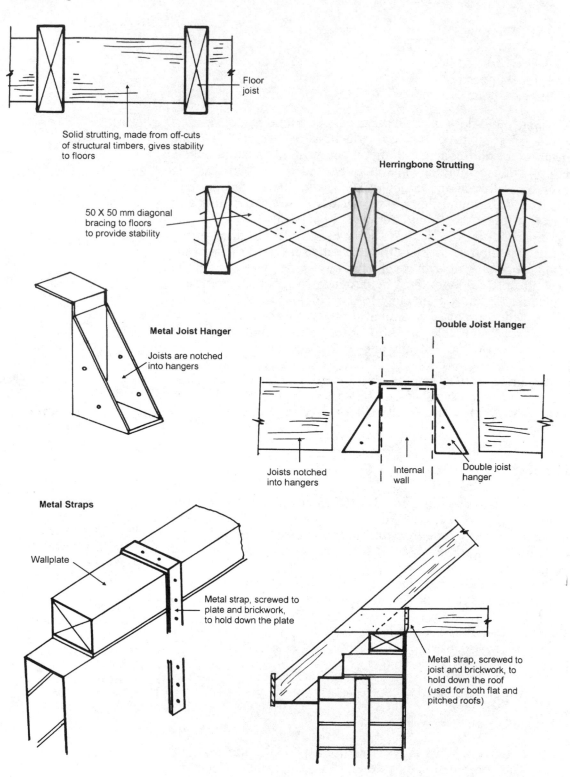

Solid Strutting

Floor joist

Solid strutting, made from off-cuts of structural timbers, gives stability to floors

Herringbone Strutting

50 X 50 mm diagonal bracing to floors to provide stability

Metal Joist Hanger

Joists are notched into hangers

Double Joist Hanger

Joists notched into hangers

Internal wall

Double joist hanger

Metal Straps

Wallplate

Metal strap, screwed to plate and brickwork, to hold down the plate

Metal strap, screwed to joist and brickwork, to hold down the roof (used for both flat and pitched roofs)

Figure 9.7 Floor and roof fixings.

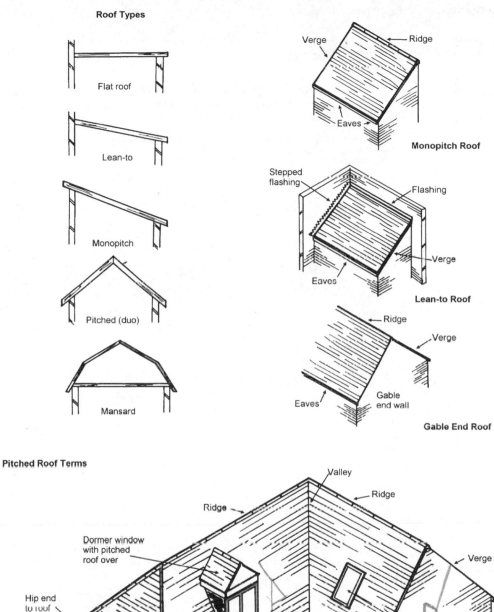

Roof Types

Flat roof

Lean-to

Monopitch

Pitched (duo)

Mansard

Verge

Ridge

Eaves

Monopitch Roof

Stepped flashing

Flashing

Verge

Eaves

Lean-to Roof

Ridge

Verge

Eaves

Gable end wall

Gable End Roof

Pitched Roof Terms

Valley

Ridge

Ridge

Dormer window with pitched roof over

Verge

Hip end to roof

Rooflight

Hip capping

Gutter

Eaves

Rainwater pipe

Figure 9.8 Roof types.

Roof terms

Rafters: sloping roof timbers which run down from the ridge to the eaves and transfer loads to the wall plate.

Wall plates: horizontal timbers along the top of a wall to which the roof beams are attached.

Ridge board: a horizontal board set on edge at the peak of a pitched roof where the rafters or other timbers meet.

Purlins: horizontal roof beams which link and brace the rafters.

Collar: a horizontal timber in tension which acts as a tie between rafters on opposite sides of a pitched roof.

Struts: timbers in compression fixed between purlins and collars in a roof structure or truss.

Ceiling joists: horizontal timbers which act as a collar and provide a fixing for the ceiling layer.

Hip: the angled meeting of two sloping roof surfaces which forms an external angle.

Valley: the meeting of two sloping roof surfaces which forms an internal angle inside a complex roof. Rain water usually drains via a valley gutter.

Gable: the upper part of an external wall where the slopes of the roof give the gable a triangular shape.

Hip end: the use of a hip to end a roof rather than a gable.

Mansard roof: a pitched roof where each face has two slopes, the lower slope being the steeper.

Eaves: the bottom edge of a roof where it usually overhangs a wall, and the area beneath this overhang.

Soffit: a horizontal board fixed to the underside of the rafters of overhanging eaves.

Fascia: a vertical board fixed to the ends of the rafters.

Typical roof structures

The headings below give details of common roof constructions in houses in the British Isles. See also Figure 9.9 for details of roof construction. Roof coverings are described in the next section. All roofs, even heavy ones, must be securely tied down to the walls of the building to avoid lifting in strong winds.

Flat roofs

The construction of a flat roof is similar to an intermediate floor except that the roof must also be waterproof, includes thermal insulation, and must resist condensation. The top of the roof or deck sits on roof beams or *joists*.

A 'flat' roof is not really flat as it must have some slope or 'fall' to allow rain water to run away and this slope is achieved by tapering timbers called *firrings*. The junctions between the roof and any *upstands*, such as walls, must be well constructed as they are a common source of roof failure.

Timber pitched roofs

If a roof surface is constructed so that it rises at an angle or *pitch* it prevents rain and snow collecting and allows roof tiles or slates to be fixed or hung.

The inclined timber *rafters* of the pitched roof rise between the wall plate and the *ridge*. Horizontal *purlins* brace the rafters, together with struts and collars.

Trussed rafters

A modern pitched roof is constructed from trussed rafters, which are prefabricated triangular frames. The rafters, ceiling ties and struts which make the frame are assembled and jointed in the factory, usually by metal connector plates.

Instead of a ridgeboard and purlins, the trusses are linked and braced together by metal straps, tiling battens and diagonal braces. This structure can be quickly erected on site by semi-skilled labour without the time and expense of a traditionally constructed roof made from cut timber.

Roof coverings

The final covering on the roof structure must keep rain water out, resist wind effects and survive the effects of heat from the

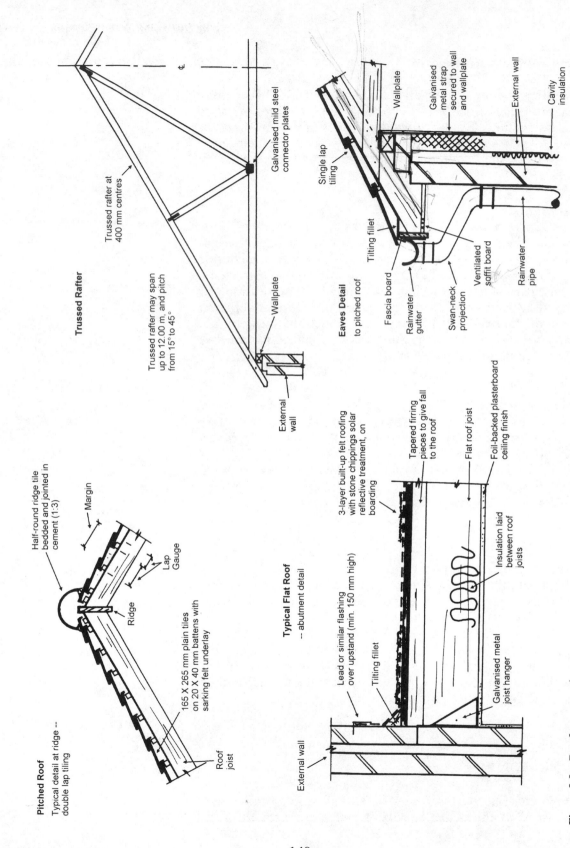

Trussed Rafter

Galvanised mild steel connector plates

Trussed rafter at 400 mm centres

Trussed rafter may span up to 12.00 m, and pitch from 15° to 45°

Wallplate

External wall

Eaves Detail
to pitched roof

Single lap tiling

Tilting fillet

Fascia board

Rainwater gutter

Swan-neck projection

Ventilated soffit board

Rainwater pipe

Wallplate

Galvanised metal strap secured to wall and wallplate

External wall

Cavity insulation

Pitched Roof

Typical detail at ridge -- double lap tiling

Half-round ridge tile bedded and jointed in cement (1:3)

Margin

Lap

Gauge

Ridge

165 X 265 mm plain tiles on 20 X 40 mm battens with sarking felt underlay

Roof joist

Typical Flat Roof
-- abutment detail

3-layer built-up felt roofing with stone chippings solar reflective treatment, on boarding

Tapered firring pieces to give fall to the roof

Flat roof joist

Foil-backed plasterboard ceiling finish

Lead or similar flashing over upstand (min. 150 mm high)

Tilting fillet

Insulation laid between roof joists

Galvanised metal joist hanger

External wall

Figure 9.9 Roof construction.

148

Sun and cold from the frost and snow. A roof system must also allow water vapour to escape from inside the roof.

Pitched roofs in the British Isles have traditionally been covered by clay tiles or slates. Flat roofs have traditionally been made waterproof with bituminous compounds, either spread directly on the roof or laid in the form of sheets.

Tiles and slates

Tiles

Clay tiles are manufactured by shaping clay and then baking in a kiln, in a similar way to making clay bricks. The tiles keep out the rain by making use of the pitch (slope) of the roof and the overlaps between each course of tiles. In a traditional double lap system each tile overlaps two other tiles. Sheets of waterproof underlay ('roofing felt' or 'sarking felt') are used beneath the tiles to complete the barrier, especially against wind-driven rain or snow.

The tiles are attached to strips of wood, called *battens*, which run horizontally between the rafters or trusses, the spacing depending on what system of tiles is to be used. At the top of each tile are protruding nibs which hang on the battens and holes for nailing the tile to the battens. Special sizes and shapes of tile are used at ends of courses, ridges, hips and eaves. Other systems of roof tiles include interlocking tiles and pantiles which have only one overlap but use special edge shapes which connect into one another.

Typical roofing materials
Clay tiles
Concrete tiles
Slates
Thatch
Copper
Zinc
Lead
Steel
Aluminium
Glass
Bituminous compounds

Roof-covering terms

Pitch: *the slope of the roof rafters in terms of angle above the horizontal. Roof pitches vary from 17.5 degrees to 45 degrees.*

Lap: *the distance by which one tile overlaps another.*

Gauge: *the distance between battens which can be calculated as*

$$\text{Gauge} = \frac{\text{tile length} - \text{lap}}{2}$$

Slates
Natural slates are produced from a form of rock which can be split easily into thin sheets and which has a distinctive dark grey 'slate' colour. Quarries in Wales and Cornwall are traditional sources. Artificial slates are composite materials using polymer matting, Portland cement and colouring pigments.

Slates are laid like double lap tiles and, as they do not have interlocking shapes or nibs, each slate must be nailed to battens.

Sheet materials

Built-up roof felts
Roofing 'felt' is made from a flexible base of fabric or polyester which is factory-impregnated with bitumen. Long strips of felt are used to 'build-up' a roof system of two or three layers. The layers are bonded together with hot bitumen and each strip is overlapped. The top surface is protected by gravel, white chippings, or reflective paint.

Profiled metal sheets
Various roofing systems use long sheets of aluminium or steel which are overlapped at their edges. A shape or profile, such as wavy corrugations, is rolled into the flat metal sheet to increase the rigidity. The steel also has a galvanised layer of zinc to help protect again corrosion.

Openings

Door and window functions
Weather resistance
Fire resistance
Security
Ventilation
Daylight inwards
Views outwards
Thermal insulation
Sound insulation
Long life
Easy maintenance

Most buildings need doors and windows which, in addition to their usefulness, have a large effect on the appearance and appeal of a building. The various functions, listed in the margin, must be balanced against one another with suitable compromises, which are not usually easy. For example, even a triple-glazed window has very poor thermal insulation compared with a modern wall; and the airtightness needed for thermal and sound insulation must be balanced against the need for ventilation.

When an opening is made in a wall, for a door or window, the structure above the opening must be supported and the loads transferred around the opening. The curve of an arch is an elegant solution but simple buildings usually use a lintel to 'bridge' the opening.

A *lintel* (sometimes spelt lintol) is a beam which gives support to the superstructure above an opening such as a window or door. Traditional lintels have been made from timber. Lintels in ready-made lengths are made from reinforced concrete and from galvanised steel which has been pressed into a box-like section and often contains insulation.

The construction of a wall opening also needs careful techniques for lining the edges of the opening, bonding the door or window frames, keeping out the weather, and maintaining the insulation.

Windows

The different areas and parts of a windows often have special names, as shown in Figure 9.10. Windows can also be described in types according to the way that they open; such as swinging outwards, sliding sideways, sliding upwards, or pivoting.

The general construction of a window can be considered under the heading of frames, and glazing as described below. The fastenings or 'ironmongery' of windows is also a significant topic, especially when security devices are included.

Window types
Casement window
Vertical sliding window
Horizontal sliding window
Louvre window
Sash window
Pivot window
Fixed window

Window frames
The outside frame of the window, which is fixed to the opening in the walls, has a *head* at the top and a *sill* at the bottom which are joined by two side *jambs*. The head and the sill usually extend from the wall so as to deflect rainwater.

The main frame may be subdivided by vertical *mullions* and horizontal *transoms* as shown in Figure 9.10. The internal *casement* or *sash frames* carry the glazing material such as glass. Timber is the traditional material for window frames but frames are also made from steel, aluminium and plastic.

Glazing material
Glass is the most commonly used material for 'filling-in' or glazing the window. Glass is usually chosen for its transparency but glass can be also be made with patterns, reinforced with wire, laminated with plastic, toughened by heat treatments, or modified to control solar radiation. *Double-glazed* units are made by sealing dry air or an inert gas in the gap between two layers of glass.

The glazing is attached to the casement frame by various methods such as clips or strips of material. The edges of the glazing must then be made waterproof by flexible materials such as putty, flexible compounds, or rubber seals.

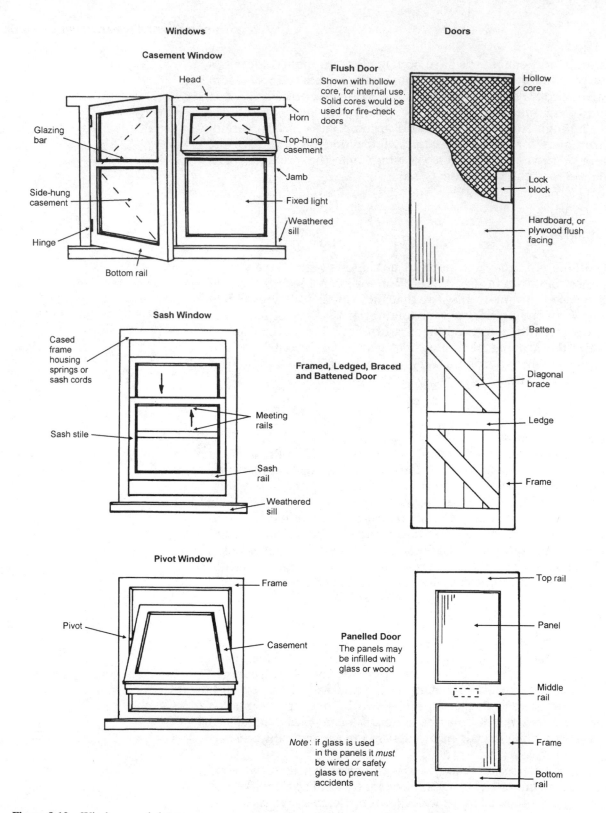

Figure 9.10 Windows and doors.

152

Doors

Simple doors are arranged to swing on hinges within a frame fixed to the wall opening. The basic components of a door frame are similar to those of a window frame: head, side posts or jambs, and sill. Other terms associated with doors are shown in Figure 9.10. Doors also need 'ironmongery', such as hinges, catches and other security devices.

The moving part of the door can be classified into various types as described below.

Door types
Framed ledged and braced
Ledged and braced
Panelled door
Flush door

Matchboard door
These are composed of a series of vertical boards held together by horizontal *ledges* and sometimes by diagonal *braces*.

Panel door
A basic frame is made from a top rail and bottom rail joined by vertical *stiles*. This frame, which may be subdivided, is filled in with plywood or glass panelling.

Flush door
A flush door has two smooth surfaces such as a thin wood veneer or a material suitable for decorating. The internal *core* of the door may be a simple timber skeleton or a cellular filling.

A fire check door is often a flush door which has been constructed to provide *fire resistance* for a designated period such as half an hour. Timber and other wood products can be used to provide above average fire resistance provided that the edges of the door are also designed to eliminate gaps.

Stairs

Stairs allow people to move between floors levels and may also be an escape route from a fire. Even the stairs in a private house are subject to design regulations to make the stairs as safe as possible by providing even, well-sized steps, handrails, headroom and maximum steepness.

Timber is the popular material for stairs in houses while concrete is preferred for flats, offices and commercial buildings. Timber stairs are usually supplied in ready-made flights to suit standard floor-to-floor dimensions.

Typical stair requirements
(for a private house)

- Equal rise for every step
- Equal going for all parallel treads
- Maximum pitch of 42 degrees to horizontal
- Going minimum of 220 mm
- Rise maximum of 220 mm
- Going plus twice rise to be 550–700 mm
- Stairway width at least 800 mm
- Headroom minimum of 2 m
- Handrail height between 840 and 1000 mm
- Balustrade openings maximum 100 mm

Key words for Chapter 9, Structures and Superstructures

The following is a list of some keywords used in this chapter. Use the list to test your knowledge and, if necessary, consult the text to learn about the terms.

Barrel vault	Flush door	Soffit
Beam	Frame	Span/depth ratio
Clay tiles	Lintel	Tie
Crosswall	Mass structure	Wall-tie
Flemish bond	Purlins	

10 Services, Appliances and Finishes

This chapter looks at the main items and finishes which go into a building after the superstructure is complete and weatherproof. There is a focus on the use of services such as water, electricity and energy supplies whose principles and arrangements have been described in earlier chapters.

Services

Water services

Every dwelling should have a piped supply of wholesome water. Chapter 5 on Services Science described the sources and treatment of water for public water supplies. This section describes arrangements within a domestic building, such as a house.

Cold water supply systems

Once inside the building the cold water can be distributed to points inside the buildings by two main methods: *direct systems* where water is taken straight 'up' from the mains, and *indirect systems* where water is taken 'down' from storage tanks, often in the roof space.

The features of the different methods of water supply are listed on page 157 and illustrated in Figure 10.1.

155

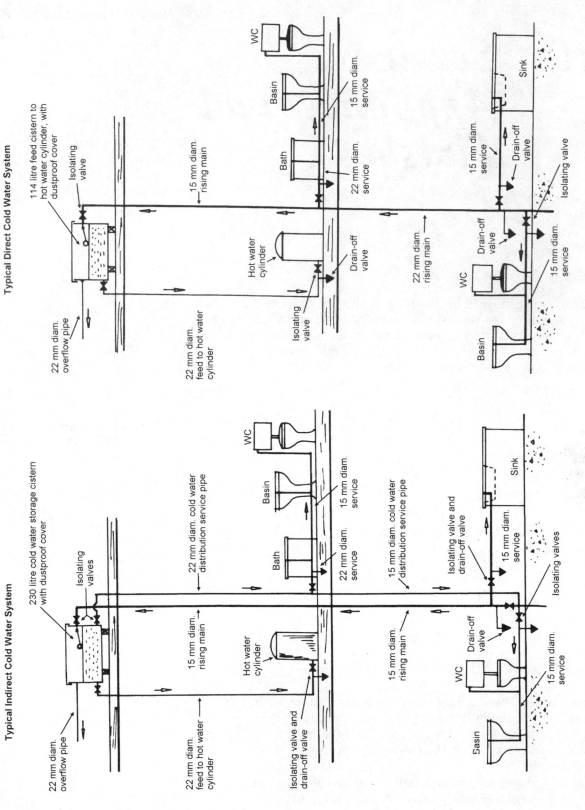

Figure 10.1 Cold water installations.

156

Indirect cold water supply
Features of indirect cold water supplies include the following:

- Water stored in cistern at higher level
- Mains is fed to storage cistern
- Cold taps not for drinking
- Demand from mains is smoothed
- Building is protected from mains failure
- More plumbing installation and higher costs.

Direct cold water supply
Features of direct cold water supplies include the following:

- No storage of water involved
- Mains supply is fed directly to all outlets
- All cold taps suitable for drinking
- Higher peak demands on mains
- Risk of back-siphonage to mains
- Less plumbing installation and lower costs.

Hot water supply systems

The supply of hot water in a building can be arranged by a number of systems and the factors affecting choice include the performance required, the fuel available and the type of heating system being used to heat the 'space' of the building. A single boiler, for example, may be able to supply both hot water and central heating.

The components of a hot water system can be arranged as a direct system or as an indirect system, in a similar manner to that described for the cold water supply. Hot water is heated and stored in a hot water cylinder with a heating system at the bottom of the cylinder. The hot water rises by convection to the top of the cylinder where it is drawn off through pipes and is replaced by cold water at the bottom of the tank.

The heating system at the bottom of the boiler may be a loop of pipes to the boiler, or heat exchanger coils containing water heated by the boiler in an endless loop of a primary circuit, or it may be an electric 'immersion' element.

Indirect hot water supply
Features of indirect hot water supplies include the following:

- Water to be heated does not pass through boiler
- Boiler is linked to hot water cylinder by a *primary circuit* which contains an endless loop of heated water

Cold water supply components
Feed cistern
Piping
Valves
Taps
Bath
Shower
Bidet
Basin
Water closet
Sink

Hot water supply components
Feed cistern
Storage cylinder
Expansion tank
Boiler
Immersion heating element
Thermostat
Piping
Valves
Taps

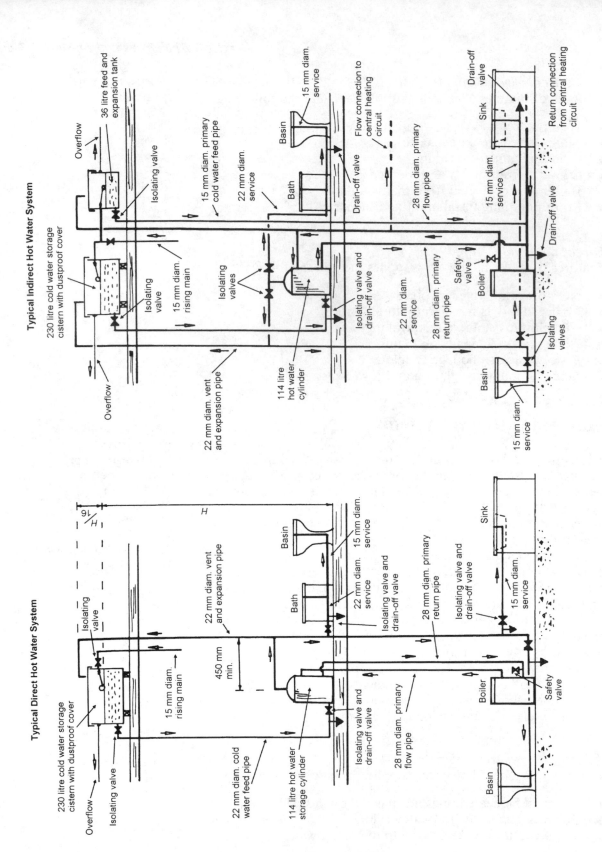

Figure 10.2 Hot water installations.

158

- Heat from primary circuit transfers by indirect contact through copper coils of a *heat exchanger*
- Suitable for hard water areas because primary circuit protects boiler from furring
- More plumbing installation and higher costs.

Direct hot water supply
Features of direct hot water supplies include the following:

- All hot water passes through boiler before storage in hot water cylinder
- Boiler prone to furring with hard water
- Less plumbing installation and lower costs.

Kitchen and bathroom fittings

The main items of equipment in kitchens and bathrooms usually require a supply of water to the appliance and a drainage system away from the appliance. We use these sanitary appliances for cleaning ourselves, for food preparation, for laundry and for disposing of waste.

Features of sanitary appliances

Sanitary appliances are usually found in bathrooms, toilets, kitchens and laundry rooms. The choice and positioning of the fittings are affected by various factors which include the following:

- Water supply: availability
- Drainage: availability
- Standardised dimensions for installation
- Cleanliness: smooth, non-absorbent, easy cleaning
- Durability: hard wearing and long life
- Ease of use: good access and design
- Economy: low initial costs and running costs
- Appearance: acceptable, fashionable
- Noise: minimised, especially in nearby rooms
- Odours: controlled, extracted
- Regulations may need to be followed.

Sanitary appliances
Water closet
Urinal
Bath
Shower
Bidet
Wash basin
Sink
Washing machine
Dishwasher
Waste disposer

Materials for sanitaryware
Fireclay
Ceramics
Vitreous enamel on cast iron or steel
Stainless steel
Plastics

The design and installation of fittings in kitchens is helped by a standardised system of measurements for storage units, worktops and equipment. This allows kitchen units and appliances from different manufacturers to be assembled into one design. Standard dish washers and washing machines, for example, all fit into a 600 mm wide space. A *modular grid* based on 600 mm squares is usual.

Regulations

Typical regulations
Building Regulations
British Standards
Water Bye-laws

The installation of sanitary appliances is often governed by national or local regulations which have been developed to help prevent loss of hygiene and dangers to health caused by poor design and installation. Some typical effects of regulations and good practice include the following:

- External windows or mechanical ventilation is required for rooms with a WC
- A wash basin should be adjacent to a WC
- Appliances must have an accessible trap which provides a water seal against odours
- Vent pipes must be installed to prevent suction breaking the water seal.

Drainage

Fresh water is brought into a building in pipes which are under pressure from the energy supplied by pumping stations. But after this water has been 'used' it leaves the building using the effect of gravity and flows 'downhill' in underground drains. There is usually no pumping of this effluent, except at specialised places in some sewerage systems.

Because the drains work by gravity flow, the effluent water can run in pipes or channels which are only part full and the jointing between parts within the drains is easier to make than joins in water supply pipes which are full of water under pressure.

Some drainage terms

Waste water: outflow of dirty water from basins, baths and washing machines.

Soil water: Outflow of human waste from WCs.

Foul water: mixtures of waste water and soil water.

Surface or storm water: rainwater or clean water collected from roofs and paved surface areas.

Effluent or sewage: the various mixtures of waste, foul and surface waters which need to be drained from a building.

Discharge pipes: the pipes that carry waste and soil and surface water to the drain.

Drain: a pipe or closed channel which carries away effluent below ground level.

Sewer: a form of drain which collects the effluent from a number of drains.

Sewerage: a network of sewers that disposes of sewage from a community.

Trap: a bend or similar device in a pipe that remains full of water and so prevents the escape of gases and smells from the drain system.
Examples: gulley, P-trap, S-trap, bottle trap.

Vent: pipes connecting a drain to the open air, with the outlet at high level to prevent odours.

The branch pipe from some sanitary appliances, such as a basin, initially does flows full bore until it joins a large vertical *stack* pipe. The stack pipe should not flow full bore but the transition from the branch to the stack may generate a syphonic effect on the water seal in the trap of the appliance. The waste water in the stack then often drops in a free fall which may cause other suction or compression effects on appliances attached to the stack. These unwanted effects can be prevented by spacing and shaping of branch pipes, by vents to the open air, or by other methods of equalising pressures in the stack.

Drainage techniques

The aims of a good drainage system include the following:

- Prevention of health risk
- Prevention of foul air entering building
- Minimum noise nuisance
- Sufficient capacity for maximum design flow
- Restriction of pressure effects such as suction and compression

Table 10.1 Drainage features and functions

Feature of drainage system	Function of feature
Traps between appliance and system	Prevents foul air entering building
Depth of water seal in trap	Helps resist suction effects from waters in stack
Vent pipes to atmospheric pressure	Helps resist suction effects from waters in stack
Fall: difference in height between points in drain	Allows water to flow under gravity Affects velocity and flow rate
Cross-section area of pipe or channel	Affects maximum flow rate, in combination with fall Affects self-cleansing
Depth of drain	Adjusts the steepness of fall Protects from surface loads such as traffic Protects from freezing and heating
Large radius bends	Helps prevent compression effects Helps prevent blockages
Junctions angled in line with direction of flow	Helps prevent blockages Helps self-cleansing
Rodding points	Access for clearing blockages in one direction
Inspection chambers	Inspection and access for clearing blockages in several directions
Bedding	Helps pipes resist forces from traffic and ground movement
Flexible joints	Helps pipes resist forces from traffic and ground movement

- Encouragement of smooth non-turbulent flow
- Self-cleansing of system by normal flows
- Minimal chances of blockage
- Easy access to all parts of system
- Long-term resistance to effects of sewage liquids and to leakage
- Protection from effects of ground pressures and ground movement
- Protection from extremes of weather
- Economic and efficient to install
- Compliance with regulations.

In order to achieve the above aims the features outlined in Table 10.1 are used in various designs of drainage systems. Some details of drainage systems are shown in Figure 10.3.

Drainage systems

The drainage system from a building usually connects to a community system and there are two main systems for public drainage: *combined* or s*eparate*. When rainwater is collected from roofs and paved areas it can be combined with the foul water from the building or it can be separately taken to a special sewer, or stream or river.

If the rainwater joins the foul water then it must be connected in such a way that high flows of rainwater do not interfere with the flows from the sanitary appliances. In contrast, there are country areas without a system of public sewers where the rainwater is usually directed straight to *soakaways* in the ground while the foul water is directed to a *septic tank*, which is like a private sewage works.

Drainage components
Rainwater hopper
Rainwater pipe
Yard drainage channel
Discharge pipe
Soil and vent pipe
Back inlet gullet
Trapped gulley
Bedding materials
Drainpipes
Joints
Rodding point
Inspection chamber
Manhole

Features of separate drainage system

- Surface water kept in separate drains and sewers from foul water
- Road contains two sewers: foul water and surface water
- Surface water remains clean and needs no treatment
- Surface water can be channelled direct to streams, rivers, sea
- Higher cost of double system for building and road
- Lower volumes and costs for foul water treatment at sewage works.

Materials for drainage
Rigid:
    Vitrified clay
    Concrete
    Cast iron
Flexible:
    Unplasticised PVC
    Reinforced plastic

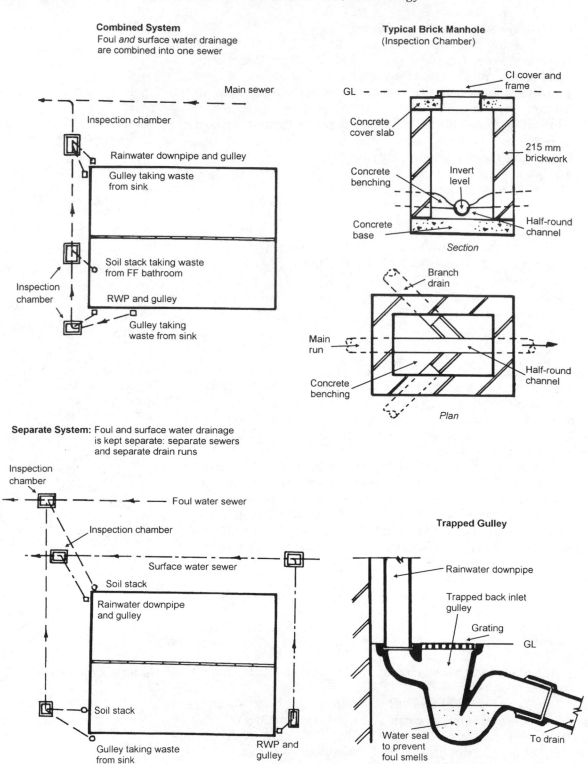

Figure 10.3   Drainage systems.

Features of combined drainage system

- Surface water shares same drains and sewers as foul water
- Road contains one combined sewer
- Surface water surges can overburden sewage works
- Cheaper and easier installation at building
- Higher volumes and costs of combined system at sewage works.

Heating systems

Chapter 4 on Environmental Science and Chapter 5 on Services Science have described mechanisms of heat transfer and how air temperature is one of the principal factors in the thermal comfort of human beings. This section outlines typical heating options for providing a comfortable temperature in a typical house or flat. Additional systems of ventilating and air conditioning are used in commercial buildings.

Energy sources
Gas
Electricity
Oil
Coal

Heating design options

If a person or a room is heated by a nearby heater while the rest of the building is cooled then the heating can be termed *local heating*. This section looks at the options for the more comfortable idea of whole-house heating in which all rooms likely to be used, are heated. The rooms might be heated by many separate heaters, such as electric storage heaters, but they usually have a single *central* source of heat such as a boiler. The boiler uses the energy source supplied to the building and converts it to heat which can be distributed around the building and reach the occupants.

Desirable targets when choosing and designing a heating system include the following:

- Environmentally acceptable energy source
- Efficient conversion of energy to heat
- Efficient distribution of heat
- Economic to install
- Economic to run
- Satisfactory appearance
- Low noise levels from equipment.

Central heating components
Water supply
Boiler
Pump
Pipework circuit
Emitter (Radiator)
Thermostat
Timer control

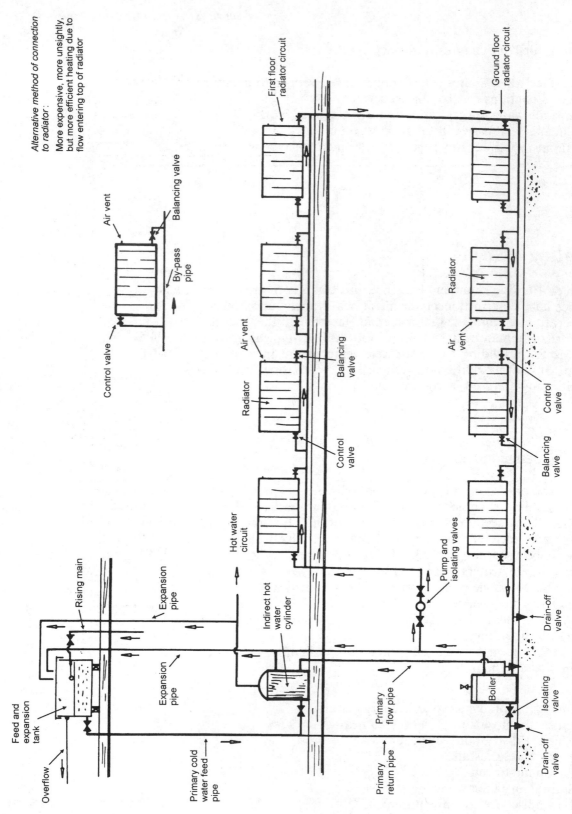

Figure 10.4 Hot water heating system – typical one-pipe system.

Alternative method of connection to radiator:

More expensive, more unsightly, but more efficient heating due to flow entering top of radiator

Air vent

Balancing valve

By-pass pipe

Control valve

First floor radiator circuit

Ground floor radiator circuit

Air vent

Balancing valve

Radiator

Control valve

Control valve

Air vent

Radiator

Balancing valve

Hot water circuit

Pump and isolating valves

Drain-off valve

Rising main

Expansion pipe

Indirect hot water cylinder

Expansion pipe

Primary flow pipe

Boiler

Feed and expansion tank

Overflow

Primary cold water feed pipe

Primary return pipe

Isolating valve

Drain-off valve

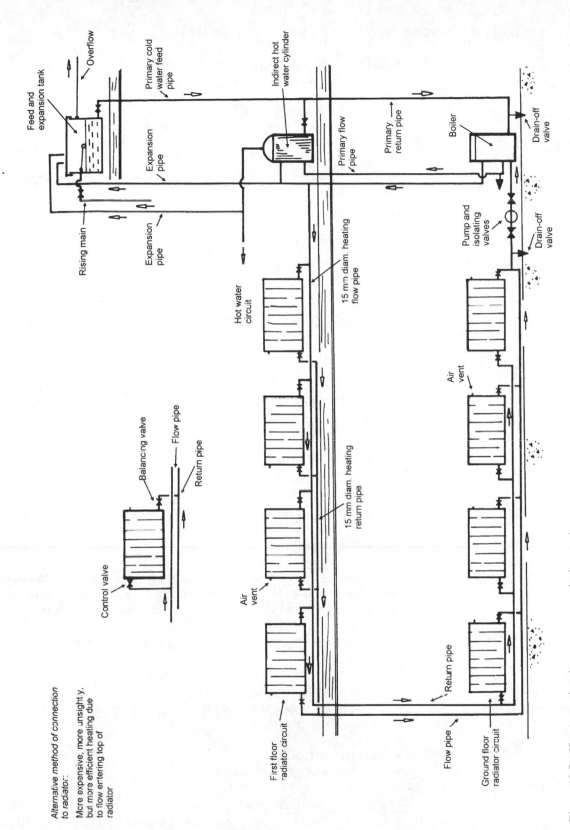

Figure 10.5 Hot water heating system – typical two-pipe system.

Feed and expansion tank

Overflow

Primary cold water feed pipe

Indirect hot water cylinder

Primary flow pipe

Primary return pipe

Boiler

Drain-off valve

Expansion pipe

Pump and isolating valves

Drain-off valve

Rising main

Expansion pipe

Hot water circuit

15 mm diam. heating flow pipe

Air vent

15 mm diam. heating return pipe

Air vent

Alternative method of connection to radiator:

More expensive, more unsightly, but more efficient heating due to flow entering top of radiator

Balancing valve

Flow pipe

Return pipe

Control valve

First floor radiator circuit

Return pipe

Flow pipe

Ground floor radiator circuit

167

Table 10.2   Technical options for heating

Performance requirement	Common technical options
Source of heat energy	Electric element Gas flame Oil flame
Distribution within building	Hot water circuits via pipes Warm air via ducts
Distribution within room	Radiators Convectors Fan convectors Grill outlets for warm air
System controls	Boiler thermostat Room thermostats Valves Time switches

Typical hot water heating systems are illustrated in Figures 10.4 and 10.5 and the function of the components are outlined in Table 10.2.

Electrical, energy and telecommunications

Electrical Supply

Chapter 5 on Services Science has described the generation and the distribution of supplies of electricity. This section outlines the arrangements for the connection and distribution of electricity in a small building such as a house.

The design aims of an electrical system in a building include the following:

- Sufficient capacity for design use
- Minimum wastage of current
- Prevention of shock
- Prevention of fire
- Means of isolation
- Compliance with regulations.

Typical electrical installations are illustrated in Figure 10.6 and the functions of the components are outlined in Table 10.3.

Electrical installation components
Service fuse
Meters
Consumer control unit
Cables
Conduits, trunking
Power circuits
Lighting circuits
Electrical appliances

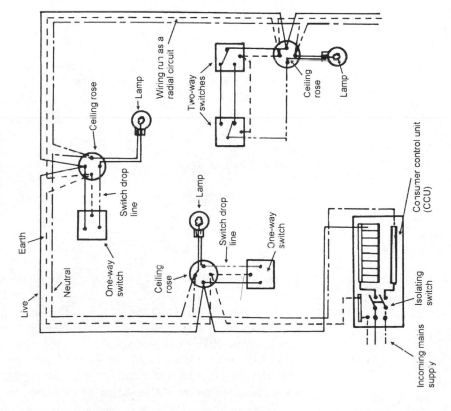

Typical Power Ring Main -- wiring diagram

Power circuits are rings. They run from the CCU and return back to the CCU at the end of the circuit

13 amp sockets on ring main

13 amp spur socket outlet

13 amp socket outlets on ring main

Live

Neutral

Earth

Isolating switch

Consumer control unit (CCU)

Incoming mains supply

Typical Lighting Circuit -- wiring diagram

Lighting circuits are radial. They run from the CCU to the last light in the circuit and stop

Ceiling rose

Lamp

Switch drop line

One-way switch

Live

Neutral

Earth

Ceiling rose

One-way switch

Switch drop line

Lamp

Wiring run as a radial circuit

Two-way switches

Ceiling rose

Lamp

Isolating switch

Consumer control unit (CCU)

Incoming mains supply

Figure 10.6 Electrical installations.

169

Table 10.3   Table of technical options for electrical supply

Requirement	Common technical options
Supply of electricity in street	Alternating current, 3-phase supply, supplied to houses in rotation
Supply of electricity in building	Alternating current, single-phase, 230 volts, taken from 3-phase street supply
Isolation between mains supply and building	Main service fuse Double pole switch
Measurement of energy consumption	Meters in house: standard tariff, low tariff
Prevention of shock	Earth connection in house Fuses, circuit breakers
Distribution to circuits	Consumer unit, bus bar
Isolation of circuits	Circuit switches, fuses, circuit breakers
Power appliances	30 amp ring circuits
Lighting provision	5 amp lighting circuits
Water heating provision	15 amp circuit
Cooking provision	30 amp cooker circuit
Conduction of current	Conductors of copper or aluminium
Prevention of energy loss in cables	Conductors of sufficient diameter
Insulation of voltage	Outer sheath of PVC, rubber
Protection of cable	Conduits or trunking of steel or plastic
Prevention of fire	Conductors of sufficient diameter Fuses, circuit breakers

Gas supplies

Forms of gas fuel
Methane
Propane
LPG (liquefied petroleum gas)

Many communities have a supply of gas for heating and cooking which is supplied by pipes in the street. The *natural gas* available in Britain is supplied by a network which brings gas from oil/gas drilling platforms in the ocean to the streets. Gas can also be supplied in storage tanks in which the gas may be kept under pressure in the form of a liquid.

Features of gas supplies

- Venting of all spaces with service pipes
- Protection of pipes by sleeve or duct
- No service pipes beneath foundations
- Flues lined with metal to prevent condensation
- Separation from electrical services.

Gas supply components
Entry to building
Isolation valve
Pressure governor
Meter
Pipework

Telecommunications

Telecommunications takes in all equipment for communicating messages between machines and people. These links may be to points within a building or to the outside. The systems used include traditional telephone systems, computer networks and satellite links (Table 10.4).

Good building design and construction make allowance for current methods of telecommunications and make it easy to adapt the building for future methods of communication.

External links
Telephones lines
Cable inlets
Satellite dishes
TV/Radio aerials

Table 10.4   Telecommunications in buildings

Requirement	Common technical options
Voice communications	Telephone lines Mobile phone networks
Document transfer	Postal services Fax telephone lines Computer networks
Information gathering	Telephone links Computer networks
Entertainment	Terrestrial TV/radio Cable TV/radio Satellite TV/radio
Fire alarms Security systems	Internal cabling Line-of-sight sensors External phone links

Mechanical transport

Lifts or other forms of mechanical transport such as escalators are usually found in commercial buildings where they give access to higher storeys or help move many people (Table 10.5).

Lift components
Shaft
Pit
Machine room
Motor
Cables
Car
Controls

Table 10.5

Mechanical equipment	Common applications
Lifts	Access between floors for people and goods Rapid access in tall buildings
Escalators	Easy access between levels, moving large numbers of people, e.g. department stores, underground railways
Travelators (horizontal)	Moving large numbers of people over horizontal distances, e.g. airports
Lightweight stair lifts	Mobility for people who cannot use stairs

Lightweight lift systems, such as those attached to the sides of stairways, can be installed in houses where they are useful for people who cannot use stairs.

Finishes

Previous chapters have described elements of the superstructure of buildings such as walls and floors. Some structural materials, such as concrete, can be left 'unfinished' but it is usual for a covering or a coating to be applied. This 'finish' or 'cladding' may be needed to protect the structural material, such as wood, and it usually needs to look satisfactory.

The finishes used in buildings and upon structures (Table 10.6) need to perform in a variety of ways, such as those listed below:

- Protection of surface beneath
- Concealment of surface beneath
- Satisfactory appearance
- Resistance to impact and abrasion
- Resistance to water
- Resistance to water vapour
- Resistance to chemicals
- Resistance to fire
- Safe to use, such as to walk upon
- Economical to buy and install
- Easy to maintain
- Satisfactory length of life.

Table 10.6 Finishing techniques

Technique	Characteristics	Typical applications
Gypsum plaster	Applied wet, by trowel, on to undersurface such as brick, block, concrete, plasterboard, insulating board	Internal wall finish with smooth hard surface Can be painted or papered
Plasterboard	Rigid building board of gypsum plaster bonded between two sheets of heavy paper Nailed on to timber battens or studs, or fixed by plaster daubs Joints may be taped and covered by skim of wet plaster or by wallpaper Board may be foil-backed if vapour resistance required	Internal finish for ceilings Internal lining for timber stud walls Internal dry lining over masonry walls
Sheet materials	Plastic surfaces, such as melamine, bonded on to sheets of paper or timber products Properties of durability and hygiene	Wall linings and fitted furniture for kitchens, bathrooms
Glazed ceramic wall tiles	Highly fired tiles with smooth impervious surface which may be coloured or patterned Applied with adhesive, with grout between tiles High resistance to liquids and to water vapour	Kitchen walls Bathrooms walls
Clay or quarry floor tiles	Fired clay tiles with hard wearing surface and high resistance to liquids Fixed by adhesive to screed or other level surface	Floors in kitchens, bathrooms, or other areas where durability is needed
Plastic floor tiles	Flexible tiles of PVC or other polymer Fixed by adhesive to screed or other level surface	Floors in kitchens, bathrooms, or other areas where durability is needed
Wallpapers	Paper which can cover under-surface and provide own decoration Types include lining paper, decorative papers, textured papers Applied with adhesive paste	Internal walls in living and bedrooms
Paint	Liquid mixture of pigment and polymer binder which dries to leave a protective and decorative film Traditional process has layers of primer, undercoat and finish coat Applied by brush, roller or spray	Internal and external components and surfaces Provides protection and decoration
Tile cladding	Clay tiles hanging vertically with overlaps Applied with nails on to timber battens with waterproof felt layer	External walls of timber framed buildings, or decoration on masonry buildings
Timber cladding	Durable horizontal timber boards in overlapping shapes such as tongue and grooved Fixed to timber frame or battens on masonry	External walls of timber framed buildings, or decoration on masonry buildings
Applied coated materials pebble dash render stucco polymer sprays	Provides durable waterproof finish to brick, block, or plywood construction which is not weatherproof Applied by trowel, machine or spray	External walls of masonry buildings or plywood clad buildings

Key words for Chapter 10, Services, Appliances and Finishes

The following is a list of some keywords used in this chapter. Use the list to test your knowledge and, if necessary, consult the text to learn about the terms.

Consumer control unit	Indirect water supply	Polymer sprays
Foul water	Local heating	Quarry tiles
Gypsum plaster	Modular grid	Separate drainage system
Heat exchanger	Natural gas	Terrestrial TV

Internal finishes
Gypsum plaster
Tiling
Paint
Paper
Sheet materials:
    plasterboard
    timber products
    plastic products

External finishes
Timber cladding
Tile hung
Metal sheeting
Applied coatings:
    paint
    pebbledash
    render
    stucco
    polymer sprays

11 *Project Development*

This chapter summarises how a development project is approached using a design point of view and a construction point of view. Common to most projects is the mobile plant of a contractor and this plant is described in the first section.

Further details of topics such as construction planning and site management are available in the companion volume: *Construction Management, Finance and Measurement* (Construction 1).

Construction plant

The selection of the appropriate site machinery or plant will help the contractor to obtain the optimum use of resources and will aid production in terms of output. The plant that a contractor uses may range from small power tools to large items of mechanical plant such as excavators and tower cranes.

Construction plant is used to bring the following benefits:

- Increase the rate of construction
- Reduce the overall costs of production
- Eliminate heavy manual work
- Replace traditional labour where there is a shortage of skills
- Maintain higher standards of quality
- Maintain the safety of personnel.

When planning to use items of plant, contractors have two main options to consider:

1. Buy the item of plant
2. Hire the item of plant.

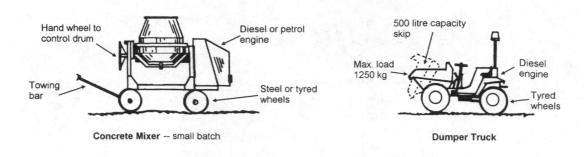

Hand wheel to control drum

Towing bar

Diesel or petrol engine

Steel or tyred wheels

Concrete Mixer -- small batch

500 litre capacity skip

Max. load 1250 kg

Diesel engine

Tyred wheels

Dumper Truck

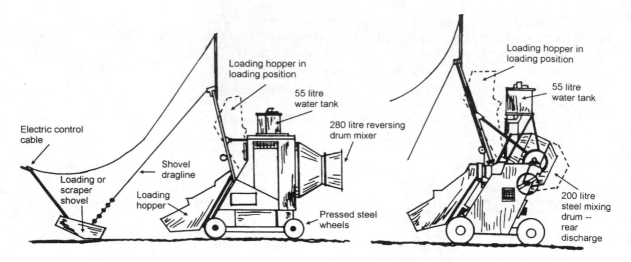

Electric control cable

Loading or scraper shovel

Shovel dragline

Loading hopper

Loading hopper in loading position

55 litre water tank

280 litre reversing drum mixer

Pressed steel wheels

Reversing Drum Concrete Mixer -- medium batch

Loading hopper in loading position

55 litre water tank

200 litre steel mixing drum -- rear discharge

Tilting Drum Concrete Mixer -- medium batch

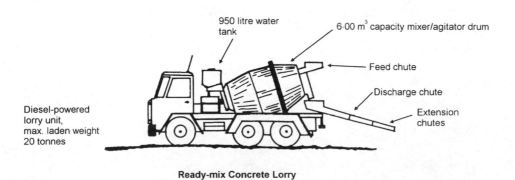

950 litre water tank

$6·00$ m^3 capacity mixer/agitator drum

Feed chute

Discharge chute

Extension chutes

Diesel-powered lorry unit, max. laden weight 20 tonnes

Ready-mix Concrete Lorry

Figure 11.1   Construction plant 1.

Generally, most smaller items of plant, which the contractor uses frequently, will be bought while larger items of plant, which are used less frequently, will be hired from plant hire companies.

Typical items of plant are illustrated in Figures 11.1, 11.2 and 11.3, and described in the following sections.

Concrete mixers

General-use mixers (or small mixers) have a rotating drum to produce a consistent concrete mix without segregation. The size of the mixer used will depend on how much concrete can be placed in a given period of time. As a guide: a batch mixing time of five minutes per cycle, or twelve cycles per hour, can be assumed as reasonable for assessing mixer output or mixer selection.

Small batch mixers are of the tilting drum type, and have outputs of up to $2-3\,m^3$ per hour, but they are usually hand-loaded so the quality control of each mix is difficult to regulate and maintain.

Medium batch mixers may be of the tilting drum or reversing drum type and have outputs of up to $10\,m^3$ per hour. These mixers usually have integral weight batching hoppers, scraper shovels and water tanks, and these features produce more consistent quality mixes.

Ready-mix concrete lorries are used to transport concrete to and around a site. The capacity of the drum is up to $6\,m^3$ and is discharged by reversing the rotation of the drum; either directly into place or else into a dumper, a crane skip or a concrete pump for final distribution. Ready-mix concrete may be either batch mixed (the lorry simply transports the concrete), or lorry mixed.

Dumper truck

Diesel- or petrol-powered dumpers have a skip capacity of up to 600 litres and are used for transporting materials around a construction site.

Dozers

Dozers (or bulldozers) are usually diesel-powered vehicles which have tracks or wheels and are used for large-scale excavation of

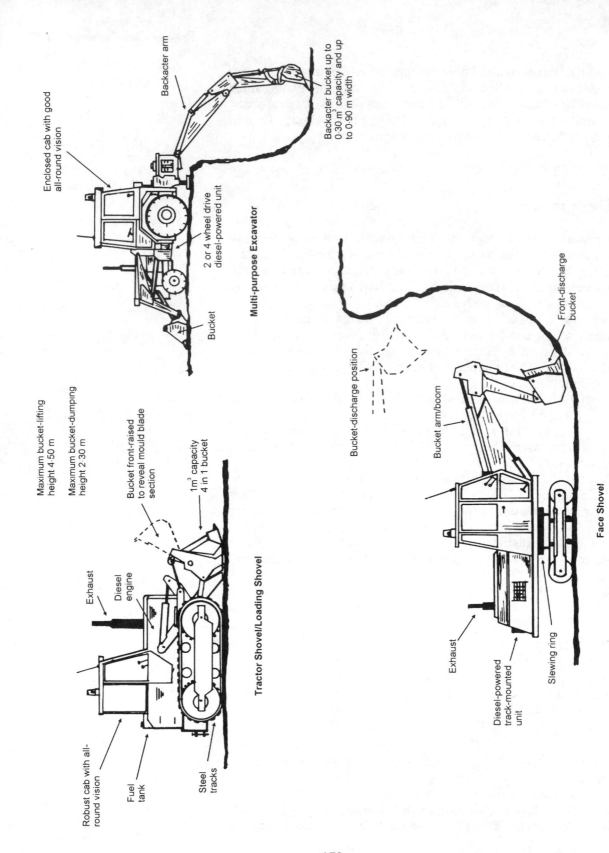

Multi-purpose Excavator

Enclosed cab with good all-round vision

Backacter arm

Backacter bucket up to 0.30 m³ capacity and up to 0.90 m width

2 or 4 wheel drive diesel-powered unit

Bucket

Tractor Shovel/Loading Shovel

Maximum bucket-lifting height 4.50 m

Maximum bucket-dumping height 2.30 m

Bucket front-raised to reveal mould blade section

1m³ capacity 4 in 1 bucket

Exhaust

Diesel engine

Robust cab with all-round vision

Fuel tank

Steel tracks

Face Shovel

Bucket-discharge position

Front-discharge bucket

Bucket arm/boom

Exhaust

Diesel-powered track-mounted unit

Slewing ring

Figure 11.2 Construction plant 2.

178

soil. The front-moulded blade can dig up to 300 mm depth and can be tilted up, down and sideways to form any required angle of slope. Bulldozers may also be used to pull other types of plant such as scrapers, rollers, scarifiers and so on.

Tractor shovels

Tractor shovels are sometimes called *loader shovels* and their primary function is to scoop up loose material and deposit it into lorries or trucks. They are usually tracked or wheeled vehicles. Tractor shovels may be fitted with a 4-in-1 bucket for bulldozing, excavating, lifting and loading operations.

4-in-1 bucket operations
Digging
Loading
Scraping
Fork lift

Multi-purpose excavators

Often known as a 'JCB' after a popular make, these excavators are usually diesel-powered with either a two-wheel or a four-wheel drive and they are used for smaller excavation works. The front bucket is often a 4-in-1 bucket system with a bucket capacity of up to $1 m^3$ and a width of up to 2 m. The 'backacter' version, used for digging trenches, has a bucket capacity of up to $0.3 m^3$, a width of up to 0.9 m and an excavating depth of up to 3.6 m.

Face shovel

Face shovels are diesel-powered units with wheels or tracks. They are used for large-scale excavation of all soils, except hard rock, and can be used for excavations up to 0.4 m in depth below their track base.

Hoists

Hoists are used for the vertical transportation of materials, passengers, or both. When in use, material hoists must be stabilised, tied into the building, and the hoist power unit must be enclosed with a protection screen made out of scaffolding in order to protect the workforce from falling materials or debris.

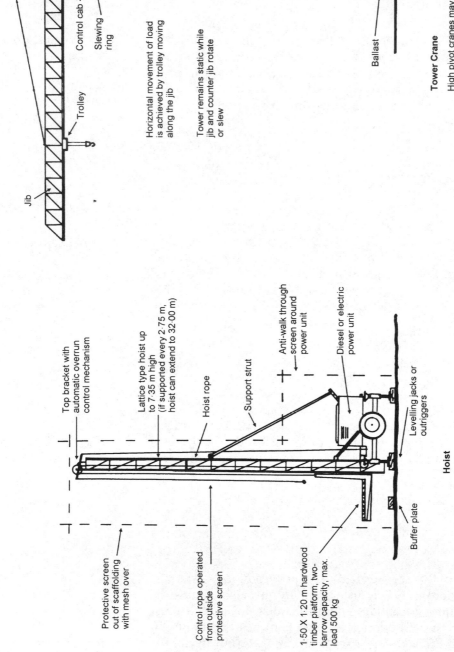

Tower Crane

High pivot cranes may be static, mobile or climbing

Counter jib

Counterweight

Tower

Access ladder to control cab

Control cab

Slewing ring

Horizontal movement of load is achieved by trolley moving along the jib

Tower remains static while jib and counter jib rotate or slew

Trolley

Jib

Ballast

Hoist

Top bracket with automatic overrun control mechanism

Lattice type hoist up to 7·35 m high (if supported every 2·75 m, hoist can extend to 32·00 m)

Hoist rope

Support strut

Anti-walk through screen around power unit

Diesel or electric power unit

Levelling jacks or outriggers

Protective screen out of scaffolding with mesh over

Control rope operated from outside protective screen

1·50 X 1·20 m hardwood timber platform, two-barrow capacity, max. load 500 kg

Buffer plate

Figure 11.3 Construction plant 3.

180

Tower cranes

Hoists are lifting devices used for raising materials both vertically and horizontally by means of steel rope operation. The range and therefore the capacity of such cranes can vary greatly. They can be grouped as mobile, static, or climbing. The correct selection of the crane depends on the nature of the project and its required workload.

Tower crane types
Mobile
Static
Lifting

Development planning

The construction of a building can be a complicated and difficult operation. Each construction project is unique and therefore requires a considerable amount of planning and preparation.

Factors which contribute to uniqueness of a building project are as follows:

- Site location
- Size
- Type of building
- Nature of site
- Materials used
- Time constraints
- Timing of construction
- Form of contract
- Construction methods
- Technology used
- Development team
- Legal requirements.

It is because of these factors that each construction project has to be considered individually and it is helpful to follow a systematic approach.

A commonly used approach or guideline is the Royal Institute of British Architects (RIBA) 'Plan of Work'. The Plan of Work sets out each stage in the development process. The stages should not be seen as mutually exclusive and do not necessarily follow one after the other. Different procurement methods may involve some of the stages occurring at the same time. For example, design and build procurement methods may have design work still continuing after construction has started.

However for the purpose of explanation each stage of the Plan of Work is described separately in the following sections.

RIBA Plan of Work

Stage	Title	Combination title
A	Inception	Briefing
B	Feasibility	
C	Outline Proposal	Sketch plans
D	Scheme Design	
E	Detailed Design	Working drawings
F	Production Information	
G	Bills of Quantities	
H	Tender Action	
J	Project Planning	Site operations
K	Operation on Site	
L	Completion	
M	Feedback	

Inception

At the inception stage a client decides to build and seeks the help of a professional consultant. It is then the responsibility of the chosen consultant to:

- Establish the client's requirements
- Visit the site
- Gather basic information in relation to the proposals
- Determine general parameters; for example, how much the client wants to spend, when the client wants the finished building, and so on.
- Prepare a brief of the client's requirements
- Advise about other consultants needed to establish the design team.

Feasibility

Under feasibility, studies will be undertaken to determine:

- Design approach
- Construction approach
- Planning issues
- Legal requirements
- Site suitability
- Project feasibility.

The consultant should either carry out a site investigation or arrange for one to be carried out. Outline planning permission should be applied for and a feasibility report prepared for the client. The site investigation should be carried out as detailed in Chapter 7 of this book.

When the feasibility report is received, the client should be able to decide, with the help of the consultants, whether to proceed with the proposal. If the client decides to proceed then the rest of the RIBA plan of work can be followed.

Outline proposals

The outline proposal stage involves the preparation of outline requirements and alternative proposals in terms of layout, design and construction of the project. Close involvement of all the consultants is required at this stage in order to assess the impact of different design solutions in relation to services, structure and aesthetics, for example, and the effect of these on the client's budget.

Scheme design

The scheme design is now worked upon to enable each component and element of the project to be designed and costed in depth. Compliance with legal requirements is taken into account and a building regulations submission is made in order to obtain approval for the scheme.

Production information

In order to get the proposal built, production information must be produced to enable a contractor to execute the works. This stage deals with the production of this information and may include:

- Production drawings
- Specification
- Schedules
- Contract details.

The production drawings, specification and schedules should be sufficiently detailed in order to allow the contractor to price the works.

Bills of Quantities

The Quantity Surveyor can use the production information in order to produce a Bill of Quantities which consists of general information about the project together with the quantities measured from the drawings in accordance with a standard measurement code, normally SMM7 or CESMM3.

The Bill of Quantities and the production information will enable the contractor to produce a tender for the project.

Tender Action

At the tender action stage a list of contractors to be selected is prepared and invitations to tender are sent out. If the contractors express a willingness to tender, the tender documents can then be issued with a date for return clearly indicated.

The following documents should be issued to the selected contractors:

- Bills of Quantities
- Project drawings
- Form of tender
- Envelopes for return of tender
- Covering letter.

Project planning

At the project planning stage the building contract for the project is placed with a contractor and discussions regarding construction work take place. The contractor can now commence planning the works in detail, placing contracts with subcontractors and suppliers, and generally arranging the site set-up and site organisation.

Operations on site

The project is now underway on site and constant monitoring and control of the operations take place. Communications at this stage are important and close collaboration between the design team and the contractor is vital to ensure a successful outcome for the client.

Completion

On completion of the project (except for snagging works) the contractor should be issued with a certificate of practical completion and the building handed over to the client.

Feedback

This stage involves the final account settlement between the contractor and the client. Records should be kept of costs and experience gained about construction methods and site organisation for future reference.

Construction planning

The above section illustrates a systematic approach to the whole development process. Similarly, a systematic approach should be taken by the contractor when considering the construction phase of the project.

A useful checklist for construction planning may include the following items:

- Information
- Resources
- Relations
- Integration of subcontractors
- Quality
- Time
- Cost
- Health and Safety.

Information

The contractor should obtain all necessary information in relation to the project. This information will include:

- Site visit report
- Working drawings
- Specification
- Bills of Quantities
- Legal obligations
- Schedules.

Schedule types
Windows
Doors
Ironmongery
Glazing
Finishes
Decorations

The information is required to allow the contractor to plan and organise the operations with optimum use of resources.

Resources

All resources for the project need consideration: labour, materials, plant, finance and management skills. The correct balance and approach to the construction process can be decided when considering the availability of labour, the type of plant and the financial control systems to be used.

Site facilities also require consideration as the correct selection and siting of facilities can go a long way towards creating an efficient and effective operating site.

Site facilities
Toilets
Storage
Canteen
Offices

Relations

The correct approach to relations with the labour force and the general public will help mitigate against unnecessary conflict and possible loss of production time. The contractor needs to consider the site management structure, as shown in Chapter 3, and the system to be used for dealing with communications on site.

The provision of good welfare facilities and proper equipment with respect to health and safety will also help to ensure a contented workforce. Good relations with the general public is basic good site practice. Notices informing the locals about the site works and possibly holding open meetings with local residents are examples of good public relations.

Integration of subcontractors

The management of subcontractors and their integration with each other and the main contractor is a prerequisite for successful construction. Decisions have to be made as to whether subcontractors are to be labour-only subcontractors, or labour and materials, for example. It is necessary to establish what facilities the main contractor will need to provide for the subcontractors and who is responsible for the upkeep of these facilities. Previous experience of the main contractor in working with a particular subcontractor will prove invaluable.

Quality management

Systems should be in place which will ensure that the client receives the specified quality of materials and workmanship. Sampling and testing procedures, together with quality checklists, should be used and their importance should be communicated to the workforce.

Time

A contractor may achieve a competitive edge over competitors by timing the works correctly. The sequencing of the operations should be programmed to present a logical approach and avoid any conflict or clashes between activities. The programme should allow the optimum use of labour and plant to avoid idle time and interruption of the works.

A typical list of operations for the construction of a simple house is given on page 188.

Cost

The programme for the works can be used as a basis for cost projections and proper financial management. The projected cash-flow compared with the actual cash-flow will enable the contractor to monitor and control the project in terms of finance. Poor management of cash-flow is one of the main reasons for construction company failures.

Collection of information which will enable actual cost to be monitored is an important function of the site management team.

Health and safety

The contractor must make every effort to comply with current health and safety legislation. The safe approach to construction methods and procedures will have an effect on the programming of the works.

The following is a list of some typical health and safety legislation affecting construction:

- Health and Safety at Work Act
- Management of Health and Safety at Work Regulations

Quality checklist for materials
Specification
Scheduling
Requisitioning
Ordering
Receiving
Handling
Storage
Security
Issuing
Incorporating

Programming techniques
Networks
Line of balance
Bar charts/Gantt charts

- Manual Handling Operations Regulations
- Personal Protective Equipment at Work Regulations
- Noise at Work Regulations
- Control of Substances Hazardous to Health Regulations
- Control of Asbestos at Work Regulations
- Control of Lead at Work Regulations
- Construction, Design and Management Regulations

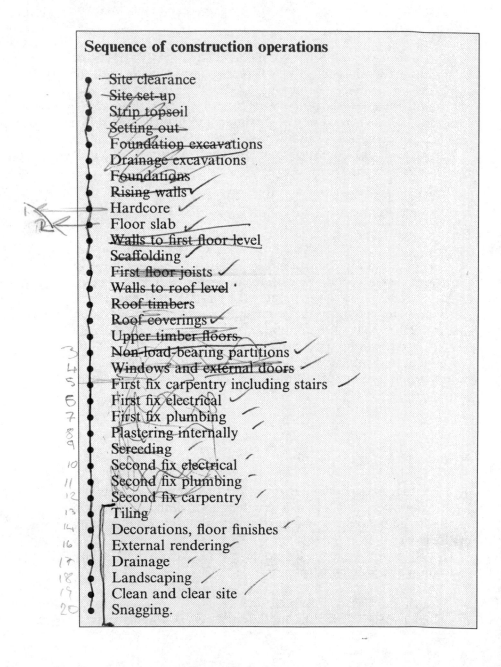

Sequence of construction operations

- Site clearance
- Site set-up
- Strip topsoil
- Setting out
- Foundation excavations
- Drainage excavations
- Foundations
- Rising walls
- Hardcore
- Floor slab
- Walls to first floor level
- Scaffolding
- First floor joists
- Walls to roof level
- Roof timbers
- Roof coverings
- Upper timber floors
- Non-load-bearing partitions
- Windows and external doors
- First fix carpentry including stairs
- First fix electrical
- First fix plumbing
- Plastering internally
- Screeding
- Second fix electrical
- Second fix plumbing
- Second fix carpentry
- Tiling
- Decorations, floor finishes
- External rendering
- Drainage
- Landscaping
- Clean and clear site
- Snagging.

Key words for Chapter 11, Project Development

The following is a list of some keywords used in this chapter. Use the list to test your knowledge and, if necessary, consult the text to learn about the terms.

Dozers	Hoist	Resources
Face shovel	Plan of Work	Scheme design
Feasibility	Relations	Sequencing
Health and Safety at Work Act		

Index